DROUGHT IN BULGA

T0249989

REPUBLIC OF BULGARIA

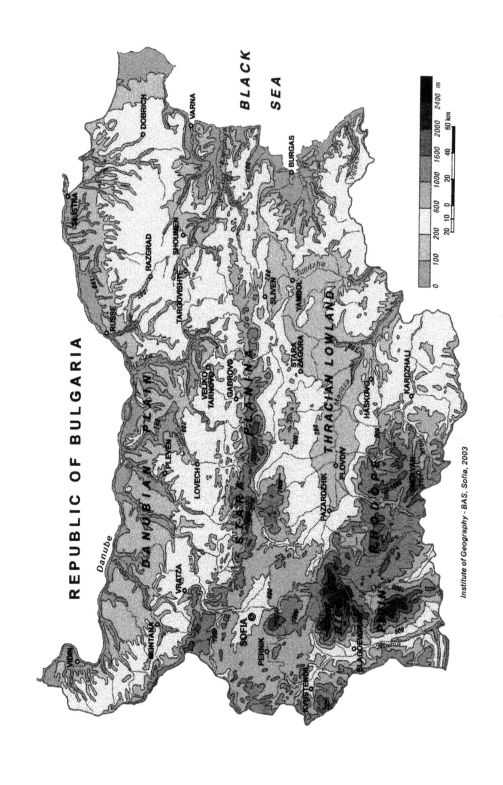

Institute of Geography - BAS, Sofia, 2003

Drought in Bulgaria
A Contemporary Analog for Climate Change

Edited by

C. GREGORY KNIGHT
Pennsylvania State University, USA

IVAN RAEV
Bulgarian Academy of Sciences, Bulgaria

MARIETA P. STANEVA
Pennsylvania State University, USA

Routledge
Taylor & Francis Group

LONDON AND NEW YORK

First published 2004 by Ashgate Publishing

Reissued 2018 by Routledge
2 Park Square, Milton Park, Abingdon, Oxon, OX14 4RN
711 Third Avenue, New York, NY 10017, USA

Routledge is an imprint of the Taylor & Francis Group, an informa business

First issued in paperback 2018

A Library of Congress record exists under LC control number: 2004014740

ISBN 13: 978-0-815-38866-1 (hbk)
ISBN 13: 978-1-138-62207-4 (pbk)
ISBN 13: 978-1-351-12660-1 (ebk)

Contents

List of Contributors

Vesselin Alexandrov (Ph.D.), Senior Researcher at the National Institute of Meteorology and Hydrology, Bulgarian Academy of Sciences (BAS), Head of the Climatology and Meteorology Networks Division. Author of many publications in the field of agroclimatology and agrometeorology including climate change and simulation models.

Galia Bardarska (Ph.D.), Senior Researcher at the Institute of Water Problems (BAS). Research on effective water treatment technologies through the use of Bulgarian filtration material and reagents. Author of a number of projects for sustainable management of water systems. Member of the Council for Central and Eastern Europe in the Global Water Partnership.

Elena Bojilova (Dr., Eng.), Research Scientist at the National Institute of Meteorology and Hydrology (BAS). Hydrological analysis and forecasts, disaggregated modeling, mathematical ecology and ecological economy.

Nikolay Chkorev (Ph.D.), Senior Researcher at the Institute of Economy (BAS), Regional and Sectoral Economy. Engaged in economic activities of environmental protection, poverty theories, regional economy.

Orlin Dekov (Ph.D.), Institute for Plant Protection, Kostinbrod. Specialist in mammal pest control in agricultural systems. Author of publications and creator of several inventions.

Ilian Dimitrov (Ph.D.), Researcher, Environmental Hygiene Section, National Center for Hygiene, Medical Ecology and Nutrition. Many years of research in water quality with special emphasis on pesticide pollution of water, the study and assessment of health risks, and epidemiology.

Hristo Dobrev (Chemical Engineer), President of SOLVO Ltd. Inventor of a number of new products for inorganic synthesis including coagulant for water purification. His reagents, patented in Bulgaria and abroad, have been used in water treatment plants for drinking water since 1976.

Georgi Fotev (Ph.D.), Professor of Sociology and Director of the Institute of Sociology (BAS) 1990-2003. Minister of Science and Public Education (1990). Member of the European Academy of Sciences and the Arts, the International Academy of Informatization, as well as other national and international academic institutions and organizations.

Marin Genev (Ph.D.), Senior Researcher at the National Institute of Meteorology and Hydrology (BAS). Specialist in general hydrology and engineering hydrology, water resources, hydrological forecasts, global climate change, ecology and protection of the environment, and artificial lakes.

Strahil Gerassimov (Ph.D.), Professor in the National Institute of Meteorology and Hydrology (BAS). Specialist in general and engineering hydrology; hydrometrics; hydrological networks, analysis and estimates; forecasts, water resources and natural changes; anthropogenic factors; flooding; drought.

Galina Gopina (Ph.D.), Senior Researcher in the National Center for Hygiene, Medical Ecology and Nutrition. Research interests in the field of negative impact of chemical pollutants and microbiological pollutants, toxicology, norms for chemical pollution of waters, evaluation of exposure to pollution of drinking water, evaluation of health risks, and epidemiological surveys.

Milena Gospodinova (Biologist), Institute of Zoology (BAS), Biology and Ecology of Terrestrial Species Section. Specialist in population morphology and diversity of mammals. Author of several publications.

Todor Hristov (Ph.D.), Professor and former Director of the Institute of Water Problems and former Secretary, National Coordination Center for Global Change (BAS). Specialist in hydro-technical structures, geo-mechanics, numerical methods, use of water resources.

Veska Kambourova (Ph.D.), Researcher at the National Center for Hygiene, Medical Ecology and Nutrition. Researcher in the field of water quality, pollution of drinking water and assessment of health risks, epidemiological studies, modeling and forecasts of exposure of the population to water ecosystems.

C. Gregory Knight (Ph.D.), Professor of Geography at the Pennsylvania State University (USA). Expert in environmental management and water resources. Founding director of the Center for Integrated Regional Assessment at Penn State. Active in the development of global change activities in Bulgaria and central and eastern Europe.

Maria Kocheva (Biologist), Researcher in the Biology and Ecology Section, Institute of Zoology (BAS), on terrestrial species. Specialist in population morphology and diversity of mammals.

Ekaterina Koleva (Ph.D.), Senior Researcher in the Meteorology Department of the National Institute of Meteorology and Hydrology (BAS). Author of many publications on climatology, data processing, climate change.

Georgi Markov (Ph.D.), Professor in the Institute of Zoology (BAS). Expert in population biology and evolution of mammals; publications in the field of population-biological and genetic-taxonomic studies of mammals.

Nadejda Nikolova (Ph.D.), Professor and former Director of the Institute of Water Problems (BAS). Researcher in the field of hydrology, modeling and regulation of runoff.

Rossitsa Nikolova (Hydroengineer), National Movement "Ekoglasnost" and Centre for Social Practices. Head of the Water Resources Department in the National Water Council (1992-1997) and head of the Strategies and Policy for Integrated Water Resources Department in the Ministry of Environment and Water (1997-2000).

Tatiana Orehova (Ph.D.), Former hydrogeologist at the National Institute of Meteorology and Hydrology (BAS), now senior researcher in the Institute of Geology. Researcher in the fields of dynamics of subterranean waters, monitoring and regime of subterranean waters, and mathematical modeling of geo-filtration.

Grigor Penev, Chief Hunting Specialist at the Ministry of Agriculture and Forests. Director of projects for sustainable exploitation and protection of biological resources and fauna.

Ivan Raev (Ph.D.), Professor and Director of the Forest Research Institute (BAS) 1995-2003. Specialist in forest hydrology and climatology. Chairman of the Committee for Forestry at the Council of Ministers of the Republic of Bulgaria, 1991-1992. President, Scientific Coordination Center for Global Change (BAS).

Boyan Rosnev (Ph.D.), Professor in Forest Protection and Forest Phytopathology at the Forest Research Institute (BAS). Author of many publications in the field of forest phytopathology, protection from biotic and abiotic damage, forest selection and ecology.

Nicola Slavov (Ph.D.), Professor of Agricultural Climatology and former Head of the Agrometeorology Section in the National Institute of Meteorology and Hydrology (BAS). Author of many publications in the field of agrometeorology, agroclimatology, and climate change.

Caedmon Staddon (Ph.D.), Senior Lecturer in Human Geography in the School of Geography and Environmental Management, University of the West of England (UK). His Ph.D. in geography at the University of Kentucky (USA) focused on the Sofia water crisis of 1994-1995. His research interests emphasize political geography and political economy of resource management in nations in transition.

Marieta P. Staneva (Ph.D.), faculty member in the Mathematics and Natural Sciences faculty at Penn State Altoona, a regional campus of the Pennsylvania State University (USA). She has a professional interest in human perceptions of global change, climate change, and the changing geography of eastern Europe.

Christopher Steuer, former graduate student in geography at the Pennsylvania State University (USA). His undergraduate degree was in earth sciences, for which he completed a thesis on climate change scenarios for Bulgaria in the context of the Balkan Peninsula. He has a special interest in geographic information systems.

Stoyan Totev (Ph.D.), Senior Researcher and Deputy Director, Institute of Economics (BAS). Research interests in regional economy, countries in transition, and comparative economics.

Liem Tran (Ph.D.), Assistant Professor, Florida Atlantic University. Formerly Research Associate in the Center for Integrated Regional Assessment at the Pennsylvania State University (USA). Tran began his career as a geologist and environmental engineer in his home country of Vietnam. He completed his Ph.D. in Geography at the University of Hawaii with specialization in environmental modeling.

Stefan Tsonev (Economist), Head of the Environmental Statistics Department, National Institute of Statistics of the Republic of Bulgaria. Specialist in environmental statistics.

Kosta Vasilev (Ph.D.), Senior researcher at the Environmental Hygiene and Water Quality Section, National Center for Hygiene Medical Ecology and Nutrition. Researcher on pesticide water pollution, health risk analysis, chemical and microbiological pollutants in drinking water. Member of the team drafting water quality regulations.

Stefan Velev (Ph.D.), Senior Researcher, Institute of Geography (BAS). Author of many publications in physical geography, climatology and nature protection.

Victoria Wesner is a student at the Lewis and Clark Law School in Portland, Oregon (USA), with a strong professional commitment to environmental management. Former undergraduate honors student and research assistant in the Center for Integrated Regional Assessment at the Pennsylvania State University.

Stiliana Yancheva (Ph.D.), Senior Researcher, Institute of Water Problems (BAS). Specialist in management of water resources, and the design and implementation of mathematical models for management and distribution of water within hydro systems.

Foreword

This book provides an excellent example of international and interdisciplinary collaboration on an issue of deep concern in the 21st century. What can we learn from the drought years in Bulgaria in 1993-1994 that can help us prepare for climatic changes in the future? The book shows that we can learn a lot by combining detailed information from the past with projections for the future. The abnormally dry conditions of the early 1990s in Bulgaria could become the norm by the middle of this century, and this bears consequences for the environment, economy and society. The roles of water management and communication of scientific results to the public and to decision-makers appear to have been critical in the past and will be most important in the future.

The topics addressed in this integrated study are at the heart of research on the human dimensions of global environmental change, which looks at the causes and consequences of, as well as the responses to, change. The vulnerability of ecosystems and society to change, the institutions that contribute to change and respond to it, the linkages between research, observation, assessment and decision-making, and issues of human security are all central to the human dimensions research agenda. The innovative methodology used in the study provides a strong basis for looking at human-environment interactions.

At the World Summit on Sustainable Development held in Johannesburg in 2002, there was considerable emphasis on the role of science. Science for sustainable development must be integrative across all disciplines; it must be "place-based" and it must address the human-environment system as a whole. This study is an example of such work. The challenge will be to ensure that its warnings are taken seriously by the decision-makers at local, regional and global levels. The book demonstrates clearly that future threats to environment and society as a result of climatic change are real and well-founded management decisions will be necessary. At the same time, it demonstrates the importance of looking at the evolving social, economic and cultural contexts in which global environmental change is taking place.

Finally, I would encourage scholars and practitioners who read this book to think about carrying out other similar case studies in order to improve our understanding of the challenges that sustainable development poses in the 21st century.

Jill Jäger
Former Executive Director
International Human Dimensions Programme
On Global Environmental Change (IHDP)

Vienna

Preface and Acknowledgements

This book is an outgrowth of global change activities in Bulgaria beginning in the mid-1990s. At that time, Bulgaria became a participant in the United States Department of Energy Country Studies Program. A team coordinated through Energoproekt, the planning arm of the Bulgarian National Electric Company, studied the greenhouse gas contributions of Bulgaria, selected impacts of potential climate change (particularly on forests and agriculture), and opportunities for greenhouse gas mitigation, especially through energy conservation (Simeonova *et al.* 2000). In June 1997, as part of an international global change research component sponsored by the U.S. National Science Foundation, the Center for Integrated Regional Assessment (CIRA) at the Pennsylvania State University and Bulgarian Academy of Sciences hosted a workshop on Global Change and Bulgaria held at the American University in Bulgaria in Blagoevgrad. Working papers on global change had been commissioned to survey the status of Bulgarian research on or related to global change issues. Working groups developed potential topics for on-going research, and the workshop as a whole developed a research agenda for Bulgaria as well as proposing the formation of a national body to coordinate activities. A petition to the President of the Bulgarian Academy of Sciences was accepted by the Academy's Board of Directors in July 1997, creating the National Coordination Center for Global Change (NCCGC). Subsequently the papers and discussions of the workshop were published by the NCCGC and Academy (Hristov *et al.* 1999; Knight *et al.* 2000).

During the 1997 workshop, we suggested the recent Bulgarian drought as one potential interdisciplinary case study related to global change that could contribute to the international literature on global climate change analogies, an area of increasing attention as society contemplates the eventualities of climate change. Bulgaria had recently experienced a severe drought with devastating effects (Knight *et al.* 1995), the culmination of a long period of increasing aridity. It was argued that if the Bulgarian economy and water use continued at the same levels as the pre-1989 political period, the drought would have brought national disaster. Such a crisis was avoided only because irrigation and industrial water demands had sharply diminished. Still, Sofia's main water supply reservoir was nearly emptied (see Hristov and colleagues, this volume);

severe water rationing in Sofia occurred widely (see Fotev, this volume); proposals for increased construction of interbasin water transfers resulted in civil disobedience (see Staddon, this volume); and eventually the national water planning system was reorganized.

Given climate change scenarios that indicate the potential for a warmer and drier Bulgaria (see Chapter 1), it was suggested that events of the 1994-1995 drought had the possibility of more frequent occurrence. Thus careful retrospective analysis of the drought, its impacts, and responses could provide both an important analogy for climate change in the country and equally meaningful lessons about policy and management response to future climate and climate variability. This suggestion was endorsed by the workshop participants and further developed in collaboration between CIRA and NCCGC, who in turn nominated Raev to direct the project from the Bulgarian side.

Operational leadership for the project was assumed by Raev, with Knight coordinating the conceptual framework, financial support, and review processes, the latter drawing upon CIRA colleagues as well as Bulgarian reviewers. Raev led the selection of lead and contributing authors and convened meetings of the Bulgarian contributors. After internal review by co-authors, the resulting chapters of what would become this book were further reviewed in Bulgaria and the US and subsequently put into final form. Raev assumed editing and publication coordination in Bulgarian (Raev, Knight and Staneva 2003); Knight and Staneva assumed similar editorial roles for this revised publication in English.

During the research that led to this book, Bulgaria assumed an increasingly important role in climate change research in central and eastern Europe (CEE; Knight 1998). Representatives of NCCGC participated in exploratory planning activities for CEE participation in global change and industrial transformation research held at the International Institute of Applied Systems Analysis in 1998. Subsequently the NCCGC was invited to participate in the formalization of the International Human Dimensions Programme's (IHDP) Industrial Transformation Science Plan (Staneva and Vassilev 1999). The NCCGC was a co-sponsor of the 1999 Budapest Workshop on Integrated Regional Assessment of Climate Change (Jäger and Knight 1999; Knight and Jäger 1999). In June 2000, a workshop on Human Dimensions of Global Change research was held by NCCGC in Sofia, initiating Bulgaria's National Human Dimensions Committee under NCCGC and IHDP auspices. In 2003, the NCCGC was reorganized as the Scientific Coordination Center for Global Change in the Bulgarian Academy of Sciences.

It is important that we acknowledge the many sources of support as our work has proceeded. Financial support for this work came to the NCCGC via CIRA from the U.S. National Science Foundation program on Human Dimensions of Global Change (Grant SBR-95-21952). In-kind support came from the NCCGC and the Forest Research Institute in the Bulgarian Academy of Sciences, as well as from the institutions represented by the various authors. We also thank Galia Bardarska (Bulgaria) and Laura Carnes (USA) who undertook major editorial and/or publication responsibilities.

C. Gregory Knight, Ivan Raev, Marieta P. Staneva

References

Hristov, Todor, C. Gregory Knight, Dimitar Mishev, Marieta Staneva (editors). 1999. Глобалните промени и България. Sofia: National Coordination Center for Global Change, Bulgarian Academy of Sciences.

Jäger, J. and C. G. Knight. 1999. "Integrated Regional Assessment in Central/Eastern Europe," *START Network News* 5:9-10.

Knight, C. G., S. P. Velev, M. P. Staneva. 1995. "The Emerging Water Crisis in Bulgaria," *GeoJournal* 35(4): 415-423.

Knight, C. G. 1998. "The Challenge of Global Change for Bulgaria," pp. 479-487 in *Papers of the 100th Anniversary of Geography at Sofia University.* Sofia: St. Kliment Ohridski Press.

Knight, C. G. and J. Jäger. 1999. "Integrated Regional Assessment of Climate Change: START CEES Workshop," *IHDP Update, Newsletter of the International Human Dimensions Programme on Global Environmental Change,* Number 4/99.

Knight, C. Gregory, Todor N. Hristov, Marieta Staneva, Dimitar Mishev (editors). 2000. *Global Change and Bulgaria.* Sofia and University Park: National Coordination Center for Global Change, Bulgarian Academy of Sciences and Center for Integrated Regional Assessment, Pennsylvania State University.

Raev, I., C. G. Knight, M. P. Staneva. 2003. Засушаването в България, съвременен аналог за климатични промени. София: Българска академия на науките.

Simeonova, K., C. Hristov, C. Vassilev, S. Todorova. 2000. "Greenhouse Gas Emissions and Mitigation," pp. 11-54 in Knight *et al.* 2000.

Staneva, M. P., C. Vassilev. 1999 "Bulgaria–A Case Study," pp. 1-25 in *National Case Studies and Regional Overviews.* Amsterdam: IHDP and Free University.

PART I
THE DROUGHT

Chapter 1

Introduction

C. Gregory Knight, Ivan Raev, Marieta P. Staneva

Introduction

From the early 1980s through the mid-1990s, Bulgaria experienced an increasingly severe drought, culminating in record-breaking proportions in 1993 and 1994. At the same time, computer models developed by atmospheric scientists to simulate climate changes resulting from greenhouse warming suggested that Bulgaria would, on average, become warmer and drier.

This book brings these two themes together—the drought of the 1990s and the potential for climate change in Bulgaria. What if Bulgaria's future climate were to be more like the conditions of the dry 1990s than today's "normal" climate? What could we learn from the impact of the drought and responses to it that could inform policies for adaptation to future climate change? Could we learn something about potential planning and responses, and, at the same time, increase awareness of the importance of greenhouse gas mitigation that might decrease the climate changes that the atmospheric scientists predict? These are the questions we set about to address.

In this chapter, we discuss the evolution of concern about climate change and how this concern eventually reached the Bulgarian academic community. Then, we turn to the questions about the drought posed to a group of scientists, largely Bulgarian but also of international origin (USA, UK, Vietnam). The broad concept of the book is that of an analogy, a kind of inference from one case study that could inform another case that is related in some way. Here, the constant is Bulgaria as a geographic locale and society; the variables are from the real, experienced drought of the 1990s versus future climate. We use analysis of what occurred in the period 1982-1994 to make inferences about planning for future climate in the same country. We do so on the basis of reasonable evidence that future climate may be like, or even worse, than the drought Bulgaria experienced.

Climate Change

Although the threat of climate change due to human emissions of greenhouse gases emerged as an important scientific and public policy issue in the 1970s, the roots of human understanding of greenhouse warming are nearly two centuries old (Clark *et al.* 2001). The greenhouse effect itself, the absorption and re-emission of long-wave (heat) radiation by atmospheric gases, was described in France by Fourier (1824); four decades later in the United Kingdom, Tyndall (1862) pointed out the role of carbon dioxide and water vapor as such gases; and in Sweden, three decades hence, Arrhenius (1896) wrote that coal burning and the resulting carbon dioxide could lead to global warming. In the 1930s and 1940s, Callendar (UK) further investigated CO_2 impacts on climate (Callendar 1938, 1949), and Plass (1956, USA) again brought CO_2 warming to the attention of the scientific community at the time foundations were being laid for the development of numerical simulation of global weather (now known as atmospheric general circulation models, or GCMs; Edwards 2000).

Several seminal studies in the 1970s suggested the importance of human-induced climate change, including two studies preparatory to the 1972 United Nations Conference on the Human Environment in Stockholm: the Study of Critical Environmental Problems (1970) and the Study of Man's Impact on Climate (1971). In 1974, Russian scientist Budyko was among the first scholars to focus on the potential of a comparatively quick change of climate caused by human economic development (Budyko 1974). This idea was supported by evidence of increased atmospheric CO_2 from numerous measurements, such as those of the Mauna Loa observatory (Keeling *et al.* 1976; *Trends '93* 1994). The Intergovernmental Panel on Climate Change was instituted in 1988 as a consensus voice of the global scientific community on climate change issues (IPCC 1992).

In 1992, at the United Nations Conference on Environment and Development in Rio de Janeiro, the Framework Convention on Climate Change (FCCC) was adopted, addressing the concern of many scientists that climate change from the impact of the greenhouse gases poses a threat to environment and society. Subsequent meetings ("Conference of Parties") suggested increasingly stringent commitments for industrialized nations to mitigate greenhouse gas emissions.

Parallel to formation of the FCCC, in 1994 the United States Department of Energy announced a major international initiative, the Country Studies Program. The goals of the program were to carry out an inventory of CO_2 sources on earth; create and apply models of climate change; identify vulnerable regions; outline strategies for adaptation of ecosystems; prepare a strategy for energy efficiency; and apply measures for the reduction of emissions of CO_2, methane and other greenhouse gases. Bulgaria was among the eventual 55 participants in the Country Studies Program. The first results published in Bulgaria related to vulnerability and adaptation of forestry and agriculture under climate change, as well as other sectors (Raev *et al.* 1995; Grozev *et al.* 1996; etc.). The results were also included in the *First* and *Second National Communications on Global Change* to the FCCC (1996, 1998) and were distributed as official documents and reports in Bulgaria. The Council of Ministers of Bulgaria approved the *National Climate Change Action Plan* (1999) in July 2000 which requires all ministries and organizations to include projects for greater adaptation of the national economy to climate change.

The Drought

The aim of this book is to gather data from the drought period of 1982-1994 and pass them to future generations, noting the impacts on natural systems and on economic and social processes in Bulgaria. The working strategy of the authors is that through the study of the consequences of the long drought in the late 20th century, we could anticipate similar effects from future climate change in Bulgaria in the 21st century. Therefore, at the end of the book we offer recommendations for Bulgarian policymakers, so that society might plan for drought as a more common occurrence and adjust the Bulgarian economy and resource management toward an expected warmer and drier future. Naturally the process of drafting a suitable strategy for each field of human activity before the threshold of climatic change is a continuous process, which is only beginning.

To study the drought in detail, the researchers set themselves several questions to guide their work:

1. What were the physical, environmental processes and events contributing to the drought?
2. What biophysical consequences can be documented—both bio-ecological and hydrologic impacts?
3. What were the human impacts of the drought—both direct climate-related effects and indirect effects through environmental and biophysical consequences?
4. What was the human response to the drought—from the population, from leadership, from the policy community?

5.   How did drought move from impacts to crisis?
6.   Finally, what messages can we give to decision-makers about planning for a future climate in which such events could become more common?

Box 1.1 provides a short synopsis of conditions that lead to drought in Bulgaria. There, Velev describes the major driving forces of weather and climate in Bulgaria and the conditions that create short and long-term drought in the country. Chapter 2 is a chronology of the drought from the early 1980s through the mid-1990s. We end our analysis in 1995, when the drought broke, with continuing impacts until reservoirs were filled. We note that shorter term droughts, not as intensive as 1993 and 1994, have occurred since. In this book, the drought is investigated in much greater detail than previously described by Knight and colleagues (1995) and Velev (1996). Chapter 3 compares climate norms and the drought year of 1993 to projections of future climate in Bulgaria, establishing that *average* future climate may be as dry as, or *drier* than, the drought.

The strategy of research and presentation continues with a discussion of the nature of analogy as a method of inference (Chapter 4). Then, the drought itself is explored in three chapters in Part 3. Chapter 5 provides meteorological details of drought; Chapter 6 sets the 1990s drought in the context of a century of drought occurrence; and Chapter 7 shows how drought in Bulgaria is related to synoptic conditions in the atmosphere over Europe. The question of how drought affected water resources is addressed in Part 4. Chapter 8 documents the impact of drought on hydrological conditions, and Chapter 9 suggests ways in which drought affected water quality. Subsequently, in Part 5, drought impacts on forests and fauna are analyzed (Chapters 10 and 11). Then we turn in Part 6 to drought-related issues in managed ecosystems—planted forests, agricultural production, and crop pests (Chapters 12, 13 and 14, respectively). Part 7 turns to the human dimensions of the drought. Chapter 15 places the drought in the broader context of the sociology and ethics of water in Bulgarian society. Economic impacts are addressed in Chapter 16 and health impacts in Chapter 17.

The consequences of drought changed from impacts to crisis during 1993-1995. This was the time Sofia's major water supply reservoir was almost emptied; severe water rationing was occurring in the capital; and a dispute over water diversions to Sofia turned ugly. These events are described in Part 8. Chapter 18 documents issues of water management, particularly with regard to Sofia. Chapter 19 examines the way the media treated the water issue. And finally in Chapter 20, the dispute over derivation of water to Sofia is addressed as a crisis in the expectations of a new democracy emerging from Communist Party rule.

Chapter 21 summarizes what this team of researchers has learned from analysis of the drought, both with regard to the drought itself and concerning the data needs and methodologies for approaching drought as an analogy for climate change. The book concludes with recommendations for policymakers in Chapter 22. What could be done differently in future drought, and how can understanding of the drought inform resource planning under conditions of climate change?

The Drought and Future Climate

Is it preposterous to suggest that future climate could be more like the peak drought years than the normal climate of 20th century Bulgaria? We think not. Detailed future research may point to higher and earlier runoff in winter and spring, and extremely dry summer and autumn conditions (Chang *et al.* 2002). Some impacts of the drier annual conditions may be diminished by wise water management. Nevertheless, it is realistic to see the mid-1990s drought as a harbinger of the future.

This book provides a waymarker in the path of global change research in Bulgaria. It is our hope that this book reaches the wider global change research and policy community. We believe the emerging drought in Bulgaria has lessons beyond its borders, both for management (or lack thereof) in the case of severe drought and as a contribution to analogies for climate change. Our deepest hope is that global society acts promptly to prevent Bulgarian drought from becoming a persistent reality.

References

Arrhenius, S. 1896. "On the Influence of Carbonic Acid in the Air upon the Temperature of the Ground," *Philosophical Magazine and Journal of Science* 41:237-276.

Budyko, M. I. 1974. Изменения климата. Ленинград: Гидрометеоиздат.

Callendar, G. S. 1949. "Can Carbon Dioxide Influence Climate?," *Weather* 4:310-314.

Chang, H., C. G. Knight, M. P. Staneva, D. Kostov. 2002. "Water Resource Impacts of Climate Change in Southwestern Bulgaria," *GeoJournal* 57:115-124.

Clark, W. C., J. Jäger, J. Cavender-Bares, N. M. Dickson. 2001. "Acid Rain, Ozone Depletion, and Climate Change: An Historical Overview," pp. 22-55 in The Social Learning Group, *Learning to Manage Global Environmental Risks*. Cambridge MA: MIT Press.

Edwards, P. N. (editor). 2000. A History of Atmospheric General Circulation Models, American Institute of Physics, Center for the History of Physics. (http://www.aip.org/history/sloan/gcm/intro.html).

First National Communication on Climate Change. 1996. Bulgarian Ministry of the Environment. Sofia: Pensoft.

Fourier, J-B-J. 1824. "Remarques generales sur les temperatures du globe terrestre et des espaces planetaires," *Annales de Chimie et de Physique* Ser. 2, 27:136-167.

Grozev, O., V. Alexandrov and I. Raev. 1996. "Vulnerability and Adaptation Assessments of Forest Vegetation in Bulgaria." pp. 374-383 in J. Smith *et al.* (editors) *Adapting to Climate Change – Assessments and Issues,* New York: Springer.

Hristov, T., C. G. Knight, D. Mishev, M. Staneva (editors). 1999. *Глобалните промени и България.* NCCGC-BAS.

Intergovernmental Panel on Climate Change (IPPC), 1992. *Climate Change: the IPPC 1990-1992 Assessments* WMO, UNEP.

Keeling, C. D., R. B. Bacastow, A. E. Bainbridge, C. A. Ekdahl, Jr., P. R. Guenther, L. S. Waterman, and J. F. S. Chin. 1976. "Atmospheric Carbon Dioxide Variations at Mauna Loa Observatory, Hawaii," *Tellus* 28(6):538-51.

Knight, C. G. 1998. "The Challenge of Global Change for Bulgaria," pp. 479-487 in *Papers of the100th Anniversary of Geography at Sofia University.* Sofia: St. Kliment Ohridski Press.

Knight, C. G., S. Velev, and M. Staneva. 1995. "The Emerging Water Crisis in Bulgaria.," *GeoJournal* 35(4):415-423.

Knight, C. G., T. Hristov, M. Staneva, D. Mishev (editors). 2000. *Global Change and Bulgaria..* Sofia: NCCGC-BAS and University Park: Center for Integrated Regional Assessment.

National Climate Change Action Plan. 1999. Bulgarian Ministry of the Environment, Energoproekt, Sofia.

Plass, G. N. 1956. "The Carbon Dioxide Theory of Climatic Change," *Tellus* 8:140-154.

Raev, I., O. Grozev, and V. Alexandrov. 1995. Проблемът за бъдещите климатични промени и противоерозионните залесявания в България, pp. 84-93 in *Proceedings of the Conference 90 Years Combating Soil Erosion in Bulgaria.* Sofia.

Second National Communication on Climate Change, 1998. Bulgarian Ministry of the Environment, Energoproekt. Sofia.

Study of Critical Environmental Problems. 1970. *Man's Impact on the Global Environment: Assessment and Recommendations for Action.* Cambridge, MA: MIT Press.

Study of Man's Impact on Climate. 1971. *Inadvertent Climate Modification.* Cambridge, MA: MIT Press.

Trends '93. 1994. *Trends '93, A Compendium of Data on Global Change.* Carbon Dioxide Information Analysis Center, Oak Ridge National Laboratory.

Tyndall, J. 1862. "On Radiation Through the Earth's Atmosphere," *Philosophical Magazine* 4(25):200-206.

Box 1.1   Bulgaria's Climate

The Republic of Bulgaria is situated in the northeastern part of the Balkan Peninsula between latitudes 41°14' and 44°13' north and longitudes 22°21' and 28°36' east. Its climate is transitional between the typical oceanic climate of western Europe and the typical continental climate of the Eurasian land mass. Proximity to the Mediterranean Sea and mountainous relief add to the diversity of climatic characteristics. A significant element contributing to this diversity is the transitional nature of climate from temperate-continental in northern areas to continental-Mediterranean in the southern and eastern border areas.

The activity of the main atmospheric pressure centers and the location of the climatic fronts over the Balkan Peninsula have a clear-cut annual regime. In spring and summer the polar climatic front lies to the north of the Balkan Peninsula and tropical air masses prevail. In autumn and winter, the polar front reaches the latitude of Crete, and mid-latitude air masses are predominant. Atmospheric moisture depends on the routes of the Icelandic cyclones in spring and summer months (maximum frequency in May and June) or on Mediterranean cyclones in autumn and winter (maximum frequency in November and December). The May-June precipitation peak, associated with the Icelandic cyclones, is characteristic of the temperate-continental climate and is most marked in North Bulgaria and in the high valleys of west-central Bulgaria. The November-December maximum, which is a distinct feature of the Mediterranean climate, is observed in the southern border areas—in the southern Struma and Mesta river valleys, in the East Rhodope Mountains, in the Sakar-Strandzha district and on the Black Sea coast. The precipitation minimum in the region with temperate-continental climate occurs in February-March, while in the region of continental-Mediterranean climate it is recorded in August-September.

The Mediterranean minimum is secondary in all the remaining areas of the country. It is chiefly caused by the tropical air prevailing during these months over the Balkan Peninsula. August and September are the months with most frequent cases of short-term drought, having the longest average duration (according to Bulgarian meteorological definition, drought is a rainless period of at least ten successive days and nights). The eastern and southern border regions of the country experience droughts throughout the warm half-year period (May-September) where the average annual number of droughts is five or six. In all other lowland regions there are four or five droughts annually and in the mountain zones, from two to four. In the peripheral eastern and southern regions the droughts are the longest (average duration 16-18 days); in the rest of the non-mountain area this length is less than 15 days and in the mountains is about 10-12 days.

The transitional nature of Bulgaria's climate presupposes fluctuations in humidity. Most commonly these fluctuations are due to changes in general atmospheric circulation. Summer droughts are considered to be a normal climatic phenomenon associated with the Mediterranean precipitation minimum. They are most frequent in the second half of the summer and during the autumn and once in every two or three years the rainless periods may last for 40-50 days. Irrespective of this "characteristic" drought, each more or less persistent change in the frequency or the trajectory of the Atlantic and Mediterranean cyclones results in droughts during the other seasons—a process which affects either isolated regions or the whole country. Such changes in circulation can also be regarded as constant characteristics of Bulgaria's climate. In certain years and periods the Atlantic cyclones move on trajectories lying further north than normal. In such cases blocking anticyclones develop and remain over the territory of Central Europe and the Balkan Peninsula, leading to a rainfall decrease in May and June, severely damaging agricultural output. The advance of Mediterranean cyclones along the route from the Adriatic Sea across the Dinaric Alps towards the Panonian Lowland of Hungary and Romania causes droughts in autumn and winter which diminish the moisture content in soils and also have a negative effect on crop yields. A similar effect is exerted by the decreased number of Mediterranean cyclones. In certain years the Azores High forms persistent crests over southern Europe, merging with the East-European High. These "linkages" act as blocking anticyclones and also produce long drought throughout the country (see Chapters 5 and 7). The combination of several variations in atmospheric circulation over southeastern Europe culminates in prolonged droughts in Bulgaria which can last for a year and sometimes even longer, as we saw in 1993-1994.

Droughts are a characteristic feature of Bulgaria's climate. Although of highly varying duration and frequency, they strongly affect the natural environment, as well as Bulgarian life and activity.

Stefan Velev

Chapter 2

Chronology of the Drought

Ivan Raev, Marieta P. Staneva, C. Gregory Knight

Events during the 1982-1994 drought period show many negative impacts of unfavorable climate in Bulgaria, with environmental, economic and social dimensions. Many of these impacts multiplied in the most severe drought period of 1993-1994, continuing into early 1995. These impacts suggest expected consequences of climate change, with future warming and drying over Bulgaria. Although we do not know all impacts, we summarize data on Bulgaria for the 1982-1994 period as contributed by the chapter authors, including climatology (Slavov *et al.*, Koleva *et al.*), hydrology (Gerassimov *et al.*), water quality (Bardarska and Dobrev), forest ecosystems (Raev and Rosnev), wild mammals and birds (Markov *et al.*), agriculture (Alexandrov *et al.*), health and hygiene (Gopina *et al.*), water resources management (Hristov *et al.*), and social impacts (Fotev, Staddon), plus the editors' experiences.

1982, The Drought Begins to Emerge

Good precipitation and runoff was observed in eastern and southern Bulgaria; values were below average in northern Bulgaria. Drought began in May-June. September and October were the driest months. Moisture conditions in forests were good, as was soil humidity; timber growth was high. Wheat and maize yields were good. Construction began on five small dams (which were completed by 1988).

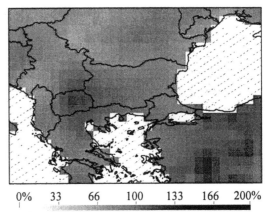

0%   33   66   100   133   166   200%

Annual Precipitation (as % of mean values)

Figure 2.1 Precipitation in 1982

1983, Drought Throughout the Year

Drought continued throughout the year. Precipitation was approximately 90% of the average; temperature was considerably below average; and runoff was about 75% of the average, with lowest values in the Danube and Black Sea basins. This was a dry year in forests, with a fall in biological productivity, an increase in the damage from pests and fungi, and drying in oak forests. Wheat yields were reduced. An epidemic outbreak of viral hepatitis A occurred among the human population. Hydroelectric power production started a downward decline. National drought planning was initiated by the Council of Ministers.

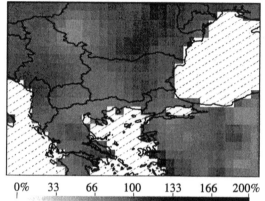

0%   33   66   100   133   166   200%

Annual Precipitation (as % of mean values)

Figure 2.2   Precipitation in 1983

1984, Drought Continues

This was the second drought year. January was dry, but there was considerable precipitation throughout February-May, then progressive drought from August to December. Precipitation was about 85% of the average. Low soil humidity, low growth of timber, continuous drought in oak forests, and outbreak of leaf-eating pests all occurred. A drastic fall of reservoir water levels was observed.

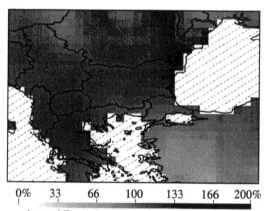

0%   33   66   100   133   166   200%

Annual Precipitation (as % of mean values)

Figure 2.3   Precipitation in 1984

1985, Serious Consequences

Drought continued in first two months of the year, and then began again from May. Precipitation was 85% of the average, with temperatures considerably above the average. Decreasing annual growth and drying in oak forests continued. Drying began in artificial lowland Scots pine and Douglas fir forests. There was an increased number of forest fires. Wheat and maize yields fell drastically. Water supply problems appeared in areas with constant shortages. A National Water Council was created. There was a marked increase in the incidence of shigellosis dysentery. Hydroelectricity production continued to decline.

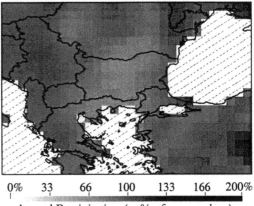

0%   33   66   100   133   166   200%

Annual Precipitation (as % of mean values)

Figure 2.4   Precipitation in 1985

1986, Ecological Impacts Continue

Drought in January was followed by a break and continued in April. Precipitation was 85% of the average, river runoff decreased to 70% of normal, and record high temperatures occurred. Drying continued in oak forests, as well as in lowland stands. Forest losses from insects and fungi increased. Sofia ignored water management recommendations developed after 1983, instead urged water diversions from the Rila region. Leaks from the Sofia water system, estimated from 40-60 percent, continued throughout the drought.

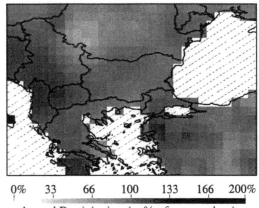

0%   33   66   100   133   166   200%

Annual Precipitation (as % of mean values)

Figure 2.5   Precipitation in 1986

1987, Continuing Environmental Impacts

Temperatures were above average, and runoff and precipitation were below average. A normally humid winter and spring was followed by drought beginning in June and continuing to the end of the year. Unsatisfactory growth in forests included greater drying in oak forests, as well as in stands of Scots pine, black pine, and Douglas fir. Drought began in lower elevation natural silver fir

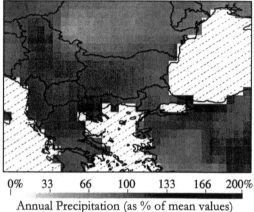

0%   33   66   100   133   166   200%

Annual Precipitation (as % of mean values)

Figure 2.6   Precipitation in 1987

forest. Unsatisfactory yields of maize point to drought effects on agriculture. Academics published a position paper objecting to water diversion projects.

1988, Water Rationing Extends

Precipitation was below average, with temperatures considerably warmer. River runoff fell to 72% of average. Increased drying occurred in oak forests, as well as in pine and Douglas fir stands. Drying in silver fir forests extended up to 1000m. Maize yields were again low. An explosion of vole populations in the country had a destructive impact on agricultural crops. Low water levels were again observed in reservoirs.

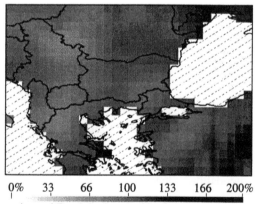

0%   33   66   100   133   166   200%

Annual Precipitation (as % of mean values)

Figure 2.7   Precipitation in 1988

Water rationing occurred in areas traditionally known for a good water supply.

1989, Water Rationing Continues

This year was again dry, with 89% of precipitation and 59% of normal river runoff. Little growth occurred in forests, accompanied by mass drying in secondary durmast and Turkey oak forests throughout Bulgaria. Maximum drying of silver fir forests was observed. Damage from forest pests and diseases increased. Shortages in water supplies in many parts of the country continued. Ecological issues contributed to public protests that led to political change late in the year.

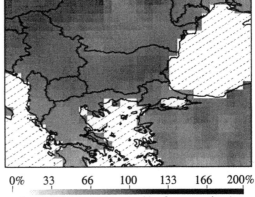

0%   33   66   100   133   166   200%

Annual Precipitation (as % of mean values)

Figure 2.8   Precipitation in 1989

1990, Drought Amidst Changes

Precipitation fell to 77% of the average, river runoff reached 46%, indicating a very dry year, with the exception of January and December. In August and September some perennial rivers went dry. Again, limited forest growth occurred, with mass drying in oak and silver fir forests. A further increase in forest fires and leaf-eating pests was noted. Maize yields

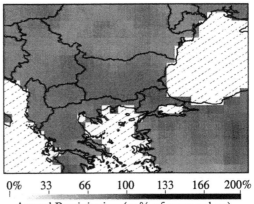

0%   33   66   100   133   166   200%

Annual Precipitation (as % of mean values)

Figure 2.9   Precipitation in 1990

again fell. Many irrigation systems collapsed due to impending land privatization. An epidemic upsurge of viral hepatitis A was observed. In response to a report from the Bulgarian Academy of Sciences, Parliament passed a decree prohibiting water diversion projects. Nevertheless, social and political changes overshadowed drought issues.

1991, Temporary Respite

A normally humid year occurred with 104% of normal precipitation and 90% of normal runoff. There was an improvement of water reserves in the soil, and fewer forest fires occurred. Drying of silver fir forests abated. There was a good wheat harvest and very good maize harvest. Unfortunately, epidemics of viral hepatitis continued, with a tendency

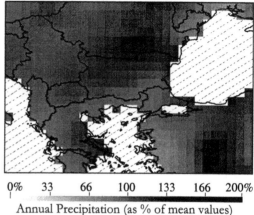

Annual Precipitation (as % of mean values)

Figure 2.10 Precipitation in 1991

towards higher incidence of shigellosis dysentery. Hydroelectric power production temporarily increased. The National Water Council was reassigned to report directly to the Council of Ministers.

1992, Drought Returns

The drought resumed with another dry year having 78% of average precipitation and 63% of river runoff. Forest growth was again reduced, accompanied by more forest fires. Once again, an increase in drying in oak forests and stands of Scots and black pine up to 700-800m was observed, along with serious drought in red oak. Some wild animal populations continued to decrease. Fall in maize yields occurred. There were minimal water levels in reservoirs, and hydroelectric power production began another multi-year decline.

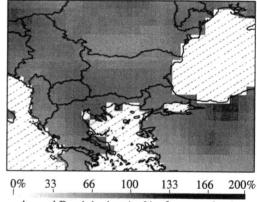

Annual Precipitation (as % of mean values)

Figure 2.11 Precipitation in 1992

1993, One of the Driest Years on Record

Precipitation was 72% and runoff 43% of normal, with record high summer temperatures. Explosive growth of forest fires occurred. Drought occurred in black locust, red oak and other stands. Drought brought record low yields of wheat and maize. In winter, interruptions in water supply occurred in some of the largest cities. Reservoir

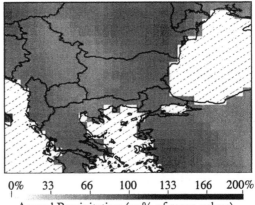

0%   33   66   100   133   166   200%

Annual Precipitation (as % of mean values)

Figure 2.12   Precipitation in 1993

depletion began, and water quality deteriorated. Hydroelectric power production again fell. National water use declined to one-third of 1989 levels. The National Water Council warned of the impending crisis.

1994, Drought Reaches Emergency Proportions

1994 was the peak of the drought: river runoff was as low as 41% of the norm. Many wild fires burned over large areas within and beyond forests. Drying of silver fir stands continued, and there was increased damage from pests and diseases in Turkey oak stands. Water shortages in many settlements continued, as well as incidences

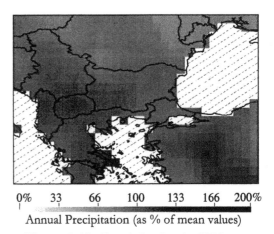

0%   33   66   100   133   166   200%

Annual Precipitation (as % of mean values)

Figure 2.13   Precipitation in 1994

of various diseases; shortages were felt virtually throughout the country. The Sofia reservoir came close to depletion, with questions about water allocations for hydroelectric production and industry instead of public consumption. A government committee formed in May imposed water rationing in September. Water prices doubled, and by the end of the year, regimes in Sofia limited water in many residential areas to as few as one day in four.[1] An increasing number of media reported on the water crisis, and public television announcements urged water conservation. Authorities prepared for medical emergencies and possible evacuations. Civil unrest emerged in the Rila region as the government revived a 1960s plan to divert water against objections of local residents and advice of many scientists, non-governmental organizations and water experts.

1995, Water Rationing until the Drought Abates

Annual precipitation returned to levels above the average, 112% of the norm. However, river runoff remained at 19% below the norm, with the lowest runoff in basins draining to the Danube River. With increased precipitation, soil moisture also improved. Abatement of forest drought could be seen, with increased growth and decreased forest

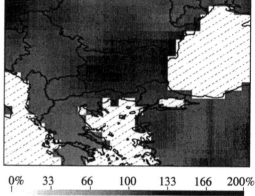

0%   33   66   100   133   166   200%

Annual Precipitation (as % of mean values)

Figure 2.14   Precipitation in 1995

fires. Wheat and maize yields also increased. Water regimes in Sofia and other cities continued into the early months. Police enforced completion of the Rila water derivation from the Skakavitsa River near Sapareva Banya to Sofia. Arguments appeared in the public media over responsibility for water loss from the Sofia reservoir. The spring thaw finally ended water shortages.

The rains came, and concerns came to an end.

Acknowledgements

The maps of annual precipitation deviation from normal (annual as a percent of the 1961-1990 mean) were provided by Liem Tran (see Chapter 7).

---

[1] In Bulgaria, water rationing is usually referred to as a "regime" in which water consumption is controlled by interrupting the supply rather than by imposition of limits on metered use.

Chapter 3

Drought and Future Climate

Christopher Steuer, C. Gregory Knight

Does the drought represent a plausible future climate in Bulgaria under existing trends of global climate change? According to studies done by the Intergovernmental Panel on Climate Change (IPCC), predicted European climate trends indicate temperature increases throughout the year, while precipitation change may fluctuate spatially and temporally (Beniston and Tol 2001). An increase in annual temperature would result in an increase in annual potential evapotranspiration (plant water needs). Increased potential evapotranspiration may result in depletion of net water supply even if precipitation increases. Periods of drought within Bulgaria, such as those experienced during 1982-1994, bring into question the stability of Bulgarian water resources given future climate change. Proper planning necessitates an understanding of whether conditions experienced during the drought years could be an analog for future Bulgarian climate.

How does possible future climate compare with the peak drought of 1993? The results of an analysis done comparing the present climate (mean climate for 1961-1990) and the drought year 1993 with climate projections lead to one conclusion: *what Bulgaria experienced in the 1993 period could well become the normal climate in future decades.*

Figure 3.1 shows the normal annual precipitation for 1961-1990 and changes expected under two climate change models for future time periods. The analysis is based on half-degree grid squares (about 40km east-west and 55km north-south at Bulgaria's latitude). In general, there may be a very small precipitation increase early in the 21st century, with small decreases by mid- and late-century. Given how dry Bulgaria was in 1993, future rainfall appears to be, on average, higher than this drought year.

However, for water resources, the issue is not just precipitation. Changes in temperature will affect evaporation. Figure 3.2 shows how future potential evapotranspiration will compare to norms. Increasing global temperatures mean higher evaporation in future, such that patterns of precipitation-minus-evaporation (Figure 3.3) indicate that Bulgaria will get both warmer and very much drier. As dry as 1993? The answer is yes, even drier.

We can map the changes in annual precipitation-minus-evaporation, comparing the future to 1993. The future compared to 1993 (Figure 3.4) shows that the future could be as dry *or drier*, over most of Bulgaria than the peak of the drought years.

These conclusions, of course, assume that greenhouse gases will continue to increase and that the models chosen represent a reasonable range of GCM scenarios of the future (they are the same models used in the United States national assessment of climate change impacts). On-going research in Bulgaria will continue this kind of analysis at a much finer spatial and temporal scale. It will be important to model water resources on a monthly or even daily basis for water planning. The technical details of how these maps were developed are included in Box 3.1.

References

Agroclimatic Atlas of Bulgaria. 1982. Агроклиматичен атлас на България. Sofia: Institute of Meteorology and Hydrology, Bulgarian Academy of Sciences.

Beniston, M. and R. S. J. Tol. 2001. "Europe," in Robert T. Watson, Marufu Zinyowera and Richard H. Moss (editors), *IPCC Special Report on the Regional Impacts of Climate Change: An Assessment of Vulnerability.* Cambridge: Cambridge University Press. Also: http://www.grida.no/climate/ipcc/regional.

Dunne T. and L. Leopold. 1978. *Water in Environmental Planning.* San Francisco CA: Freeman.

Hulme, M. *et al.* 1995. *A 1961-1995 Gridded Surface Climatology for Europe.* Climate Research Unit, University of East Anglia.

IPCC Data Distribution Centre. 2000. *Downloading Scenarios and Climate Data from the DDC.* http://ipcc-ddc.cru.uea.ac.uk/cru_data/ datadownload/download_index.html

Ivanov, N. N. 1957. Мировая карта испаряемости. Leningrad: Gidrometeoizdat.

New, M., M. Hulme, and P. D. Jones. 1999. "Representing Twentieth Century Space-Time Climate Variability. Part I: Development of a 1961-90 Mean Monthly Terrestrial Climatology," *Journal of Climate* 12:829-856.

New, M.G., M. Hulme, and P. D. Jones. 2000. "Representing Twentieth Century Space-Time Climate Variability. Part II: Development of 1901-1996 Monthly Grids of Terrestrial Surface Climate," *Journal of Climate* 13:2217-2238. Also: http://ipcc-ddc.cru.uea.ac.uk /asres/baseline/climate_baseline.html

New, M., D. Lister, M. Hulme, I. Makin. 2002. "A High-Resolution Data Set of Surface Climate over Global Land Areas," *Climate Research* 21:1-25.

Thornthwaite, C. W. and J. R. Mather. 1955. *The Water Balance.* Centerton, NJ: Laboratory of Climatology Publication 8.

U.S. Soil Conservation Service. 1970. "Irrigation Water Requirements," *U. S. Department of Agriculture Technical Release No. 21.*

Acknowledgments

The authors acknowledge the Climate Research Unit at the University of East Anglia (UK) and the Data Distribution Centre of the Intergovernmental Panel on Climate Change for access to the gridded climatology data and general circulation model scenarios.

Box 3.1 The Technical Details

Gridded values of observed monthly temperature and precipitation data for the 1993 drought year as well as mean values for the thirty-year period from 1961-1990 were obtained from the Climate Research Unit at the University of East Anglia (UEA; Hulme *et al.* 1995; New *et al.* 1999, 2000, 2002). This data has a spatial resolution of 0.5 degrees latitude and longitude. The IPCC Data Distribution Center (at UEA; IPCC 2000) provides monthly data from general circulation models (GCM) based on increased greenhouse gases and aerosols, at the grid resolutions of the respective models. For this study, the HadCM2 (The UK Hadley Centre for Climate Prediction and Research; resolution 2.5° x 3.75°) and CGCM1 (Canadian Centre for Climate Modelling and Analysis; resolution 3.75° x 3.75° over land) scenarios predicted mean temperature and precipitation differences from contemporary norms for three non-overlapping thirty-year time periods: 2010-2039, 2040-2069 and 2070-2099. These scenarios utilized the IPCC emissions scenario protocol IS92a, with sulfate aerosols and the assumption of an annual one percent increase in CO_2. The GCM data was collected and interpolated to 0.5° grid resolution using an area larger than Bulgaria to eliminate edge effects of fitting a trend surface to the changes. Temperature data at 0.5° grid resolution was used to model potential evapotranspiration (PE) using the Blaney-Criddle method (USSCS 1970; Dunne and Leopold 1978: 139-141). This method was selected among others (Thornthwaite and Mather 1955; Ivanov 1957) because it best replicated contemporary mean evapotranspiration patterns over Bulgaria (*Agroclimatic Atlas* 1982). A series of maps were then developed depicting future trends in precipitation, potential evapotranspiration, and annual precipitation minus potential evapotranspiration as an index of the humidity of climate.

Future temperature and precipitation data were segregated by GCM type and into three thirty-year intervals. Future temperature and precipitation values are reported as variations from the 1961-1990 mean for both GCMs. Temperature and precipitation data for the respective GCM grids were entered into ArcView in the form of a point theme. An Inverse Distance Weight (IDW) function interpolated these large grid cell point themes into a smaller 0.5° grid consisting of 30 rows by 40 columns. The interpolated grid sheet was designed to have a range considerably larger than that of Bulgaria to prevent edge effects in the final model. Several interpolations were performed using a range of nearest neighbors.

Point theme interpolations considering only a few neighbors produced 0.5° grid cells with little distinguishable variation. Interpolations considering all twelve neighbors were found to produce results showing the strongest differentiation between the 0.5° grid cells. The IDW considering all twelve nearest neighbors was therefore used.

Precipitation and temperature data were entered for both the observed and predicted data for each of the twelve months. Precipitation totals for each of the observed years were then calculated and reported in mm/year. Change in precipitation for future values from the observed data were calculated from the interpolated monthly data and reported in mm/year. Future interpolated changes were then added to the 1961-1990 observed values to obtain total annual precipitation values for 2010-2039, 2040-2069 and 2070-2099. Interpolated variations in temperature for each of the three future time periods were added to the mean to obtain monthly temperature means. The available monthly temperature means made it possible to calculate potential evapotranspiration using the Blaney-Criddle method, cited above. Monthly potential evapo-transpiration was calculated for each of the observed and predicted time periods. Total annual potential evapotranspiration was then summed for each of the drought years as well as the four thirty-year time periods and converted into mm/year. Changes in potential evapotranspiration were also calculated by individually subtracting the three observed time periods from the three predicted time periods.

Annual potential evapotranspiration was then subtracted from the annual precipitation totals for all time periods. The result was a series of models for both observed and future time periods depicting a precipitation-minus-potential evapotranspiration relationship. Finally, the change in precipitation-minus-potential evapotranspiration for observed years 1993 and 1961-1990 were subtracted from the predicted years of 2010-2039, 2040-2069 and 2070-2099 to obtain a series of change in precipitation-minus-potential evapotranspiration models.

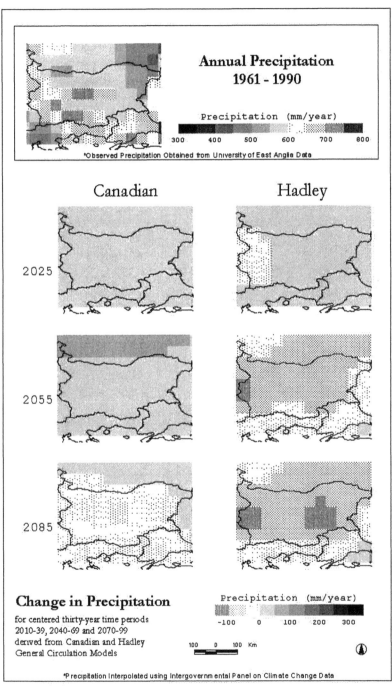

Figure 3.1 Annual precipitation

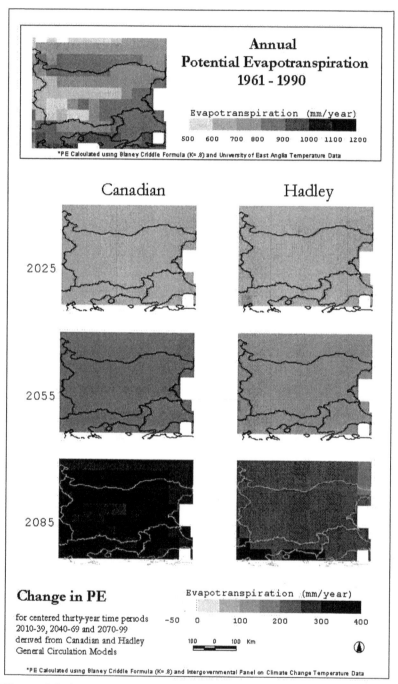

Figure 3.2 Annual potential evapotranspiration

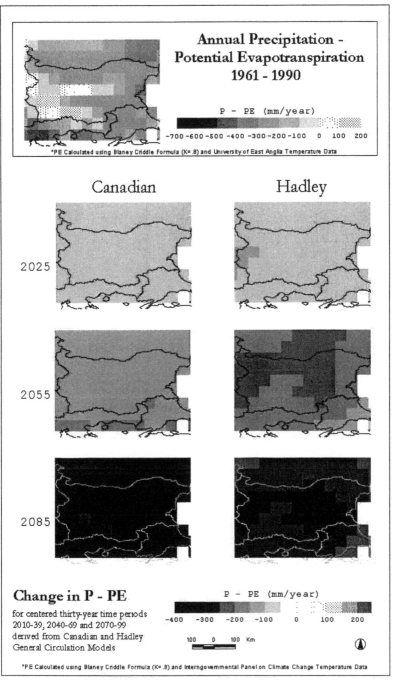

Figure 3.3 Annual precipitation – potential evapotranspiration

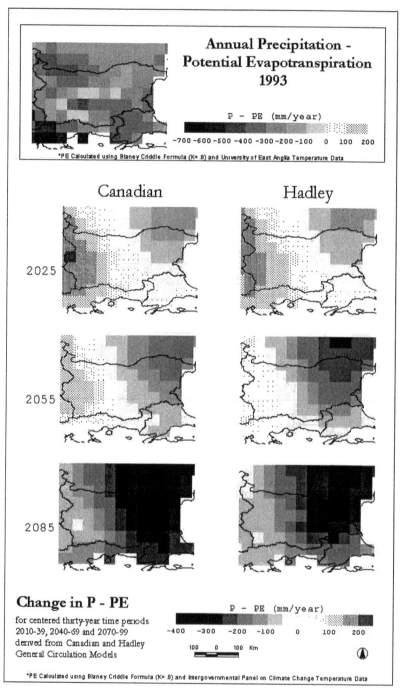

Figure 3.4 Annual precipitation – potential evapotranspiration, 1993

PART II
CLIMATE CHANGE ANALOGS

Chapter 4

Analogs of Climate Change

C. Gregory Knight, Marieta P. Staneva

This book is rooted within the "human dimensions of global change" paradigm and in the use of analogy to explore consequences of global climate change from the human viewpoint. In the ideal we would have a perfect prognostication of climate decades from now, accurate anticipation of technological and social change, and an appropriate model to link future environment and society. In reality, this is far from possible.

Analogies provide an approach to understanding the potentialities of complex events that are difficult to fully capture in formal analyses or models. Given the uncertainties in projections of future climate and the changing human context in which climate change impacts will be experienced, thinking by analogy provides one useful way to explore the dimensions of such change. A useful analogy will generate helpful questions and perspectives from an existing and well-understood event or situation that can be explored in the context of an unknown but reasonably foreseeable circumstance. The fundamental design of the research leading to this book is based on the analog concept.

Human Dimensions of Global Change

The International Human Dimensions Programme, one of the four linked global change programs (including the International Geosphere-Biosphere Program, World Climate Research Programme, and DIVERSITAS), addresses HDGC thus:

> THE HUMAN DIMENSIONS of global environmental change comprise the causes and consequences of people's individual and collective actions, including the changes which lead to modifications of the earth's physical and biological systems and affect the human quality of life and sustainable development in different parts of the world (IHDP 2000).

The emphasis on HDGC is not only on the human impacts of global change, but also on the role of human activity in creating global change (Stern *et al.* 1992). It is useful to note two kinds of global change in this context: the kinds of change that are global in distribution but not necessarily systemic (soil erosion is one example) versus change that is globally systemic such as ozone depletion and greenhouse gas warming of the atmosphere (Turner *et al.* 1991).

In the case of any spatial entity—continent, region, nation, basin—fundamental HDGC questions related to climate change include:

1. How do human activities contribute to global change, such as greenhouse gas emissions, and what are the driving forces of such human activity?
2. How do these changes affect global and regional climate?
3. How does climate change affect natural systems such as ecological systems and the hydrologic cycle?
4. How do these systemic environmental changes affect socio-economic systems such as agricultural productivity, impacts of floods or drought, and human health?
5. Which of these changes represent threats to society, and which present opportunities?
6. What options exist to adapt to threats, and at what cost?
7. What policy changes are necessary to assist in adaptation, and at a larger scale, to contribute to mitigation of activities contributing to global change?
8. How are local policy initiatives assisted or hindered by policy at the national, regional and global level?
9. Can policy changes and human actions address both the causes and consequences of global change?
10. How will alternative futures for the region differ based on actions taken in the context of processes of change and responses to existing human and environmental stresses?

The Center for Integrated Regional Assessment (USA) has captured these questions in a diagram (Figure 4.1). The diagram is an heuristic for the kinds of ideal forecasting that we know is impossible, but it suggests that human environment interaction is not a linear process but one of complex feedbacks and interactions (Knight 2000).

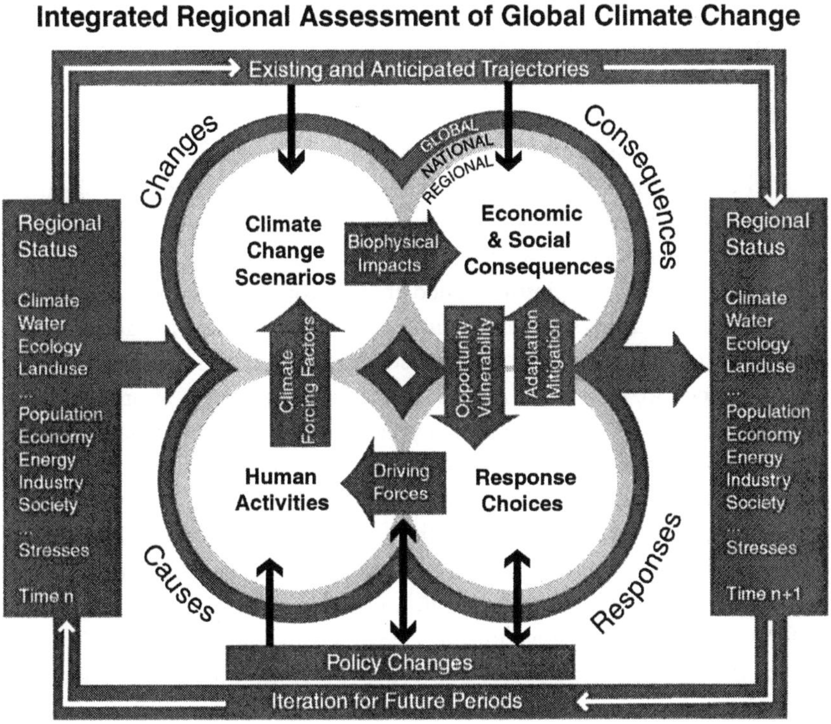

Integrated Regional Assessment of Global Climate Change

Existing and Anticipated Trajectories

Changes

Consequences

GLOBAL NATIONAL REGIONAL

| Regional Status | | Climate Change Scenarios | Biophysical Impacts | Economic & Social Consequences | | Regional Status |

Climate
Water
Ecology
Landuse
...
Population
Economy
Energy
Industry
Society

Stresses

Time n

Climate Forcing Factors

Opportunity Vulnerability

Adaptation Mitigation

Human Activities — Driving Forces — Response Choices

Causes

Responses

Climate
Water
Ecology
Landuse
...
Population
Economy
Energy
Industry
Society

Stresses

Time n+1

Policy Changes

Iteration for Future Periods

Figure 4.1 The CIRA framework for integrated regional assessment

Analogies

Glantz (1988b:3) suggested that "...analogies...are scenarios about future worlds based on human experience and as such have the political and social credibility that computer-generated scenarios lack." They provide clues and generate ideas that assist in improving understanding of phenomena. Analogies must be appropriate to the specific situation or phenomenon under study and share similarities to be most useful (Chorley 1964). Analogies can be made using a case-study (Knight 2001) or scenario approach; they can also be developed at a methodological level, using the framework or approach from another research. Examples of the former would include studies of past societies coping with climate change (Fagan 2000), and of the latter, using the insights from natural hazard research to outline the structure of possible human responses to climate change (Burton 1992; Meyer *et al.* 1998).

Analogies may be used for a variety of purposes in the case of global change. Glantz (1991) suggested that the purpose of an analogy may be as an heuristic for popular education or educating researchers; for generation of hypotheses, model parameterization, and forecasting; or for generation of policy options. Glantz (1991:14) further suggested that analogies fulfill a psychological need for understanding and a sense of control. Analogies may also be used in constructing plausible scenarios (Jamieson 1988).

The soundness of an analogy relies largely on judgment. It is insufficient to count the number of similarities, but rather a good analogy depends on "pragmatic considerations regarding contexts, interests and purposes" (Jamieson 1988:81). Jamieson argued that although *analogs* would seem to be in the nature of things, whereas *analogies* are conjectures, in fact analogs and analogies are based on analogical reasoning, a process for which there is no formal procedure (1988:81). Advantages of a case scenario approach based on analogy include incorporation of a wealth of detail; integration of a broad range of knowledge; possibility of a multiple perspectives; and communicability and usability (Jamieson 1988:82-86). The construction of the analog provides details beyond the actual forecast scenario. Moreover, understanding the past provides details which are difficult to model quantitatively, but can be included in a narrative. These details bring together different types of knowledge into forecasting. In addition, information on past events can be gathered from numerous persons, preserving each "voice" within the narrative, what Jamieson (1988:85) calls a "multiplicity of scenarios." Finally, analogs are more easily communicated and applied when presented to various audiences, particularly laypersons and policymakers.

Analogies must be used with care. Potentially similar and apparently analogous patterns may result from different forcing functions, processes or pathways. There are important differences between sustained and cumulative change versus relatively sudden and impermanent variability. Dissimilarities could make an analogy questionable, as could inappropriate extrapolations across time and scale. Whether what *could* happen is appropriate in the absence of knowing what *will* happen; whether contemporary or past societal conditions are reasonable assumptions for the future; and whether the analogy is chosen to support preconceived conclusions are additional issues. Jamieson (1988:86-90) notes three traps or pitfalls of analogs: lack of definition that limits ways of looking at a problem; straining an analogy by ignoring misanalogies; and failure of an analogy due to different technological possibilities, differences in social and political organization, and different access to information. Jamieson (1988:91) suggests that the test of analogies or any other method is "...whether or not it contributes to our understanding of the phenomena under consideration."

Analog Thinking in Climate Change Research

Prognostication of future climate change and human adaptation is fraught with difficulty. General circulation models of climate change are themselves uncertain and divergent. Whereas most scientists agree greenhouse warming is occurring (IPCC 1995), forecasting specific temperature, precipitation and spatial variations, particularly at regional-to-local scales, is not at all certain. Extrapolation from possible climate to biophysical impacts may draw upon well-established scientific principles and models, but extension from these biophysical impacts to human impacts and potential societal responses amplifies uncertainty—society and technology will certainly evolve substantially during decades-to-century climate scenarios. In the face of these challenges, the reality of climate change is such that action now cannot prevent some degree of climate change, and failure to act decisively may lead to undesirable change (IPCC 1995) that could be avoided by GHG mitigation or moderated by planning for adaptation. One potential approach to this dilemma is to build global and regional linked models that begin to address changes that chain from climate to biophysical phenomena to human impacts to potential adaptation. This approach has the potential to compound uncertainty and test credibility, although important strides have been made in some global models, such as IMAGE (Alcamo 1998), specific sectors such as water (Gleick *et al.* 2000), and limited regional settings (Warrick *et al.* 2001). Another approach is to turn to analogy, using climate analogs to assess potential impacts or using analogs of human adaptation to past climate variability or other natural hazards to understand the future. "The purpose of looking back is to determine how flexible (or rigid) societies are or have been in dealing with climate-related environmental changes" (Glantz 1988b:3-4). The use of analogs "to formulate present and future change policies is only now being recognized as an invaluable approach to global change analysis where environmental problems have a long lead time" (Leatherman 1992:17).

There are two basic types of analogs which dominate the climate change literature—historic events and trends and regional views of present and future climate. Historical climate analogs have widespread use and play an important role in understanding global warming and its regional impacts. They may be based on paleoclimate reconstruction (Glantz 1991) or historical, instrumented observations. For example, the MINK study (Easterling *et al.* 1992; Crosson and Rosenberg 1993) developed a methodology for assessing climate change impacts on the agricultural economy in Missouri-Iowa-Nebraska-Kansas based on 1931-1940 Dust Bowl era weather records. MINK thus used real rather than hypothetical climate patterns, examining potential crop production impacts under modern agricultural practices (Easterling *et al.* 1992). The strength of this approach is that analogs are built from real (not just *realistic*) climatic conditions and from existing adaptation potentials.

Since *future* societal responses to regional impacts of decades-to-centuries long climate change are unknowable, coping strategies developed in the face of current climate in other places provide one kind of spatial or geographical analogy (Meyer *et al.* 1998:220). Comparing human adaptations in regions with climates similar to the suggested changes that could occur is one strategy (Parry *et al.* 1988). Human impacts and adaptation to past climate variability and change (or to address other natural hazards such as floods, coastal erosion, and sea-level rise) may also be instructive historical analogies (Glantz and Ausubel 1988; Meyer *et al.* 1998:220). Fagan's (2000) study of the "Little Ice Age" provides an excellent analogy for the vulnerability of society to climate variability. Societal coping mechanisms employed in response to environmental events serve as an analog for potential responses to the impacts of global warming (Glantz 1988, 1991; Glantz and Price 1992; Meyer *et al.* 1998).

Meyer and colleagues (1998:238-259) cited examples from urban climate, collapse of ancient civilizations, climate crises in medieval Europe, arctic subsistence, Great Plains agriculture, Brazilian drought, and South Asian and African climate variability as examples of climate vulnerability. Glantz and colleagues drew upon analogies such as water levels in the Great Lakes and Great Salt Lake, sea level rise in South Carolina and Louisiana, drought in the Mississippi and Colorado basins, aquifer depletion in the Great Plains, and citrus freezes in Florida. From these studies, Glantz (1988) was able to highlight insights about environmental change, human impacts, and potential responses that could be applicable elsewhere.

Analogs for climate change impacts and adaptation have great utility. In addition to general education and research, analogs are useful in developing parameters for large-scale global models, for forecasting future changes, for generating policy options, for providing a range of impacts in the face of uncertainty, and in understanding potentials and policies for GHG mitigation (Glantz 1991; Klemanski and Steel 1989; Kowalok 1993; Solomon 1995).

Bulgarian Drought as an Analog

The study of events before, during, and immediately after the 1982-1994 drought in Bulgaria follows the emerging tradition of analysis by analogy in climate change research. The increasing aridity in Bulgaria from the 1980s, culminating in the most severe drought on record during 1993-1994, is suggestive of the kinds of future climate predicted for the area using general circulation models of the atmosphere (see Chapter 3). Should mean climate become generally hotter and drier, there will surely be variability around such means, but it is reasonable to suspect that drought periods, measured against contemporary climate normals, could become more frequent or persistent.

Thus, investigation of the drought as an analogy for future climate seems both appropriate and important. The members of the research team have dissected the nature and impacts of the drought in the subsequent chapters. In a later chapter we assess what has been learned about Bulgaria and about our use of the analog method. Suggestions from each of the authors contributed to the chapter on recommendations for policymakers that concludes our work.

References

Alcamo, J. *et al.* 1998. *Global Change Scenarios of the 21st Century, Results from the Image 2.1 Model.* Kidlington, Oxford UK: Elsevier Science.

Burton, I. 1992. "Regions of Resilience: An Essay On Global Warming," pp. 257-274 in in Schmandt and Clarkson 1992.

Chorley, R. J. 1964. "Geography and Analogue Theory," *Annals of the Association of American Geographers* 54:127-137.

Crosson, P. and N. J. Rosenberg. 1993. "An Overview of the MINK Study," *Climatic Change* 24:159-173.

Easterling, W. E., N. J. Rosenberg, M. S. McKenney, C.A. Jones. 1992. "An Introduction to the Methodology, the Region of Study and a Historical Analog of Climate Change," *Agricultural and Forest Meteorology* 59:3-15.

Fagan, B. L. 2000. *The Little Ice Age: How Climate Made History 1300-1850.* New York: Basic Books.

Glantz, M. H. 1988a. *Societal Responses to Regional Climatic Change.* Boulder, Westview Press.

Glantz, M. H. 1988b. "Introduction," pp. 1-8 in Glantz 1988a.

Glantz, M. H. 1991. "The Use of Analogies in Forecasting Ecological and Societal Responses to Global Warming," *Environment* 33(5):10-13, 27-33.

Glantz, M. H. 1992. "Assessing Physical and Societal Responses to Global Warming," pp. 95-112 in in Schmandt and Clarkson 1992.

Glantz, M. H. and J. H. Ausubel. 1988. "Impact Assessment by Analogy: Comparing the Impacts of the Ogallala Aquifer Depletion and CO2-Induced Climate Change," pp. 113-142 in Glantz 1988a.

Glantz, M. H. and M. F. Price. 1992. "Summary of Discussion Sessions (Regional issues i.e. science, policy, societal, regional organization, NGOs)," pp. 33-56 in M. H. Glantz (editor), *The Role of Regional Organizations in the Context of Climate Change.* Series I: Global Environmental Change, Vol. 14. NATO ASI Series. New York/Berlin: Springer-Verlag.

Gleick, P. *et al.* 2000. *Water: The Potential Consequences of Climate Vulnerability and Change.* Oakland, CA: United States Geological Survey and the Pacific Institute for the National Assessment of the Potential Consequences of Climate Variability and Change Program.

Hesse, M. B. 1966. *Models and Analogies in Science.* Notre Dame, Indiana: University of Notre Dame Press.

IHDP 2000. *International Human Dimensions Programme.* http://www.ihdp.uni-bonn.de

IPCC 1995. *IPCC Second Assessment, Climate Change 1995.* Geneva: Intergovernmental Panel on Climate Change.

Jamieson, D. 1988. "Grappling for a Glimpse of the Future," pp. 73-112 in Glantz 1988a.

Klemanski, J. S. and B. S. Steel. 1989. "Citizen Attitudes, Knowledge and Participation in Environmental Policy-Making: Michigan, Ontario, and the Case of Acid Rain," *Michigan Academician* 21(2):175-189.

Knight, C. G. 2001. "Human-Environment Relationship: Comparative Case Studies," Vol.10, pp. 7039-7045 in N. Smelser and P. Baltes (editors), *International Encyclopedia of the Social and Behavioral Sciences.* Oxford: Elsevier.

Knight, C. G. *et al.* 2000. *The CIRA Framework for Integrated Regional Assessment.* http://www.cira.psu.edu/framework.htm

Kowalok, M. E. 1993. "Common Threads: Research Lessons from Acid Rain, Ozone Depletion, and Global Warming," *Environment* 35(6):12-38.

Leatherman, S. P. 1992. "Coastal Land Loss in the Chesapeake Bay Region: An Historical Analog Approach to Global Change Analysis," pp. 17-27 in Schmandt and Clarkson 1992.

Meyer, W. B., K. W. Butzer, T. E. Downing, G. W. Wenzel, J. L. Wescoat, B. L. Turner II. 1998. "Reasoning by Analogy," pp. 218-289 in S. Raynor and E. L. Malone (editors), *Human Choice and Climate Change Vol. 3 - Tools for Policy Analysis.* Columbus: Battelle Memorial Press.

Miller, W., R. Alexander, N. Chapman, I. McKinley, J. Smellie. 1994. *Natural Analog Studies in the Geological Disposal of Radioactive Wastes.* Studies in Environmental Science 57. New York: Elsevier.

NSF 1999. *Human Dimensions of Global Change (HDGC).* Washington, DC: U. S. National Science Foundation. http://www.nsf.gov/sbe/hdgc/hdgc.htm

Parry, M. L., T. R. Carter, and N. T. Konijn (eds.). 1988. *The Impact of Climate Variations on Agriculture.* (2 volumes). Boston: Kluwer Academic Publishers.

Schmandt, J. and J. Clarkson (editors). 1992. *The Regions and Global Warming: Impacts and Response Strategies.* New York: Oxford University Press.

Solomon, B. D. 1995. "Global CO2 Emissions Trading: Early Lessons From the U. S. Acid Rain Program," *Climatic Change* 30(2):75-96.

Stern, P. C., O. R. Young, and D. Druckman (editors). 1992. *Global Environmental Change, Understanding the Human Dimensions.* Washington DC: National Academy Press.

Turner, B. *et al.* 1991. "Two Types of Global Environmental Change: Definitional and Scale Issues in Their Human Dimensions," *Global Environmental Change* 1(1):14-22.

Warrick, R. A., G. J. Kenny and J. J. Harman (editors). 2001. *The Effects of Climate Change and Variation in New Zealand: An Assessment Using the CLIMPACTS System.* International Global Change Institute (IGCI), University of Waikato, Hamilton, New Zealand.

PART III
ANALYSIS OF THE DROUGHT

Chapter 5

The Climate of Drought in Bulgaria

Nicola Slavov, Ekaterina Koleva, Vesselin Alexandrov

Drought can occur in any month of the year. Weather conditions during drought are characterized by decreased precipitation, high air temperatures, low humidity, and warm, strong winds. Long-term drought can negatively impact the water balance of plants, causing unstable crop physiological conditions and low crop yields, as well as threaten natural ecosystems and water supplies.

Investigations of drought are carried out all over the world. However, because of the complexity of this natural phenomenon, a uniform methodology for implementing drought studies has not been developed, although some indices of drought are widely used (see Chapter 6). Drought classification is one basic reason for this dearth of methodologies. Typically, there are four types of drought. *Soil drought* occurs during long-term periods without precipitation. When this type of drought occurs, soil moisture decreases considerably, and crops and natural plant communities suffer. Reduced precipitation and high air temperatures are observed during *atmospheric droughts*. Hot, dry winds are a frequent event. Because of high rates of evapotranspiration during these atmospheric conditions, the water balance of plants is disrupted. However, crop damage is most extensive during *soil-atmospheric droughts*. Long-term droughts that reduce river runoff, underground sources of water, and moisture are called *hydrologic droughts*.

Drought in Bulgaria

Drought in Bulgaria is characterized by different features during different seasons in the year. Spring drought is characterized by normal air temperatures, low humidity, and strong winds. These weather characteristics affect conditions needed for crop sowing, germination, and development. The delay of crop germination affects crop growth and development during the entire crop-growing season. Under drought conditions, pest populations increase considerably and add additional damage to crops. Spring droughts are especially typical in northwestern Bulgaria (where 40% of Bulgarian spring droughts occur) and in the coastal region of the Black Sea.

High air temperatures, low humidity, and intense transpiration characterize summer drought. Summer drought can typically be classified as soil-atmospheric drought, especially when available soil moisture in the top 1m decreases to less than 70% of the potential capacity. Agricultural crops yellow and stop growing when soil moisture decreases considerably. Crops are especially damaged when summer drought is combined with dry winds. Summer droughts are most intense and of longest duration on the coast of the Black Sea and in the Thracian Lowland of the Maritsa River basin.

Autumn drought rarely affects spring agricultural crops because these crops have been harvested. However, autumn drought does affect the tillage, sowing, and germination of winter crops. During autumn, winter crops do not develop well and are easily damaged by frost. Autumn drought is typical on the Black Sea Coast, in northeast Bulgaria, and in the Thracian Lowland.

Long-term variations of basic meteorological phenomena in Bulgaria have been amply studied. Scientific attention has been directed at precipitation sums and average annual and seasonal air temperatures. For that reason, there are some weather stations in the country with weather data for more than 100 years. The long-term variations of the non-growing period (November-March), potential crop-growing period (April-September), and actual growing seasons of winter crops (October-June) and spring crops (April-September) have also been studied. The study of drought conditions, in which precipitation is less than 1, 5, or 10mm, depending on the developmental stage of the crop, have also been the target of significant scientific interest. Some investigators characterize drought intensity according to anomalies of monthly precipitation (e.g., ±10-20% of average amounts). Insignificant drought can be characterized by conditions where precipitation is about 75% of normal amounts; moderate drought can be characterized by conditions where precipitation is 51 to 75% of normal amounts; and severe drought is typified by conditions where precipitation is less than 50% of average amounts and monthly air temperature is 1°C higher than the norm. Special attention has also been directed to droughts with more than five successive days without precipitation. A rainy day is defined as daily precipitation above 1mm.

Drought and Impact Investigations

Investigation of drought has been important in Bulgaria. One of the first comprehensive investigations of long-term variations of air temperature and precipitation in Bulgaria was carried out by Ganev and Krastanov (1949, 1951). They used data derived from various times within the period 1887-1949 from 47 weather stations, agro-meteorological newsletters, and synoptic maps, and defined drought as more than ten successive dry days. Sabeva (1968) investigated droughts between 1896 and 1960. She assumed drought to be periods in which 10 or more days passed having less than 5mm of precipitation. Droughts were interrupted when precipitation above 5mm occurred. Sabeva determined the spatial distribution of drought during the crop-growing season April-October, because droughts occurred from month to month or sometimes covered two-three months. Stefanov (1968) investigated the circulation characteristics of drought during the period 1941-1960 using topographic synoptic maps and pressure synoptic maps at the 500mb level. Five days was assumed to be the minimum duration of a drought because the duration of the natural synoptic period is between five to seven days.

The first attempt to investigate drought impacts on agricultural crops in Bulgaria was undertaken by Kirov (1934, 1948). Drought assessment included analyses of monthly and yearly precipitation rates. Drought was observed during the first 60 years of instrumental records in 1894, 1899, 1904, 1907, 1908, 1917, 1918, 1923, 1926, 1927, 1928, 1934, 1938, 1942, 1945, 1946, 1947 and 1948. The longest drought occurred in 1918, with duration of six-seven months. Anticyclonic weather was considered a major cause of drought conditions. The major centers creating anticyclones pointed out by Kirov were the Arctic and Azores atmospheric centers. According to Dimitrov (1941), droughts can be divided into two major groups: (a) winter drought—from January to April, when Bulgaria is influenced by the Siberian anticyclone and continental cold air masses passing over the country; and (b) summer drought—from July to September, when the country is influenced by the Azores anticyclone.

The first comprehensive study of the relationship between drought and crop growth, development, and yield formation was carried out by Kiryakov (1941). He reported that drought is closely connected to different growth stages such as sowing, flowering and maturity. Negative impacts of reduced precipitation were especially significant after a severe winter or when drought conditions existed until crop flowering. These conditions were observed in 1908, 1918 and 1934. Reduced precipitation, high air temperatures and hot, dry winds affected agricultural production in 1907, 1908, 1917, 1918, 1921 and 1926. According to Kiryakov (1945), severe droughts during the first half of the twentieth century were recorded in 1918, 1928, 1934, 1938 and 1945. The longest observed drought was in 1918.

A detailed study of drought impacts on agricultural ecosystems during different seasons of the year was done by Hershkovich (1968). The investigation was directed toward four basic problems: (a) weather conditions during the crop-growing season, when droughts occur, (b) drought duration during the crop-growing season, (c) drought intensity and its impact on agricultural crop yield, and (d) agricultural regions of drought occurrence. Maps of the following droughts were created: (a) winter drought, using precipitation sums from October to March; (b) spring drought, using precipitation sums during the period April-June and soil moisture in spring; (c) summer drought, where the balance of atmospheric moisture during the period June-August and soil moisture at the end of the spring and maize yield were analyzed; and (d) autumn droughts, using precipitation sums during the period September-November. Details of all these early investigations were documented by Slavov *et al.* (2003).

Drought Intensity and Duration

Drought intensity from 1931 to 1960 was also investigated by Hershkovich (1968). Most intense were the autumn-winter droughts (October-March) in 1933, 1949 and 1959 when soil water content was less than 100mm. Drought was observed also in 1934, 1943, and 1950. Spring droughts (March-May) were recorded in 1934, 1945, 1947 and 1949. These droughts significantly affected winter wheat productivity. Very intense droughts were observed across the country in 1938, 1945, 1946, 1950, 1952 and 1958. There were some years with several drought occurrences in one year, for example in 1934, 1945, 1946, 1949, 1950 and 1952. Successive dry months during one year considerably affected agricultural production. There were numerous years with both spring and summer droughts. Similarly, both summer and autumn droughts and also autumn and winter droughts were frequently observed. Drought conditions during the entire year were registered very seldom—once every ten to fifteen years. The probability of drought occurring during at least one season during a year was relatively high, almost 80%. On the other hand, the probability of two or more droughts occurring during a single year was between 30 and 40%.

Studies addressing precipitation and drought were also carried out by Koleva (1981, 1987, 1991, 1994), who studied the period 1896 to 1990. These studies were primarily directed to analyses of long-term variations in annual, seasonal, and monthly precipitation by means of 10-year averages. It was found that negative precipitation anomalies prevailed over positive precipitation anomalies, especially in summer and winter seasons. Cycles in long-term precipitation variations with duration of 2-4 years were obtained. The characteristics of droughts during the warm half of the year (April-October) were also analyzed. The longest spring droughts (which occur during April and May) were recorded during the period 1959-1968, especially in northern Bulgaria. Maximum duration dry spells in July in North Bulgaria were observed from 1943 to 1952. On the other hand, maximum duration of the dry spells for the same month in southern Bulgaria were observed between 1937 and 1946. The dry spell occurrences in August peaked during the period 1944-1953.

Koleva's studies showed that low precipitation amounts were measured during the periods 1903-1912, 1926-1935, 1941-1950 and 1959-1968. The following years had particularly low amounts of precipitation: 1907, 1945, 1958, 1983 and 1985. The driest winters were in 1930, 1949, 1976 and 1983; the driest springs—1934, 1945, 1947, 1968 and 1983; the driest summers— 1928, 1945, 1950, 1952 and 1965; and the driest autumns—1926, 1932, 1948 and 1969. The results obtained show that precipitation during the decade 1981-1990 was 10% less than average for the studied climatic period (1961-1990). Precipitation reductions were most significant during the period 1985-1995, especially during the warm half of the year (April-October). Lower precipitation during the period (1961-1990) was measured in the central part of the Danube plain and the Struma River valley where precipitation amounts were 80% less than normal. Precipitation in these regions during the warm-half of the year in the last decade of the twentieth century was less than 65% of the norm. Drought was assessed by the duration of days with precipitation amounts between 1.0mm and 5.0mm. Compared to the average duration of drought between 1931-1969, 1961-1990, as well as during the interval 1981-1990, the average duration of droughts during the last years increased by one-three days. The longest droughts were observed at the end of the summer and the beginning of the autumn. There was an increasing trend of drought duration in winter during the last years of the 20th century. The regions with less precipitation were characterized by longer drought duration.

Groups of successive months with reduced precipitation (less than 20% of the average) were also observed. Usually two or three successive months during the 20th century had precipitation amounts that were less than 50% of the norm. The year 1945 was abnormally dry—precipitation was less than average from January to November in most areas of the country. There were several successive months with reduced precipitation in 1938, 1959 and 1968. In 1985, 1990 and 1992, there were between four and six dry successive months with less than average precipitation amounts.

There were three periods during the 20th century that were characterized by long and severe drought occurrences: 1902-1913, 1942-1953, and 1982-1994. Dry years were approximately 20% of total years during the first period. Dry years during the second period increased to 40%, and dry years during the period 1982-1994 were already 50%. The driest year for the investigated period during the 20th century (1901-1996) was 1945. Among the driest years were also 1902, 1907, 1932, 1934, 1946, 1948, 1950, 1953, 1985, 1986, 1990, 1992 and 1993.

It can be seen from the interannual precipitation distribution during the periods 1961-1990, 1942-1953 and 1982-1994 that the droughts from 1942-1953 occurred at the end of summer and the beginning of autumn. Precipitation was especially reduced in winter during the period 1982-1994. The precipitation distribution during these three periods kept its moderate continental pattern.

Months or successive months with reduced precipitation were not uncommon events in Bulgaria. On average, there were about two successive months of reduced precipitation in the lower areas of the country. However, there were also some years with six or more successive dry months. Long droughts during the cold half of the year were observed in 1913, 1934, 1967, 1976 and 1983. Long droughts during the warm half of the year were recorded in 1928, 1945, 1965 and 1985. Climatically, there were three successive months with reduced precipitation. Five total months with reduced precipitation occurred throughout the year. Two to three months during the year had precipitation amounts below 50% of the average. In some years, there were five, six, or more successive dry months. In 25-30% of the years during the period 1906-1996 there were several successive months with precipitation below the normal. The year 1945 was especially dry—precipitation was below average from January through November. This drought was very long (lasting 12-15 months) when the fact that the last months in 1994 also had reduced precipitation is taken into account. There were also several successive dry months (from January to August, in some areas to October) in 1938, 1959 and 1968. In some years, monthly precipitation was only 20-30% of the average. However, these drought conditions were not observed everywhere in the country. There were between four to six successive months with reduced precipitation in 1985, 1990, 1992, 1993 and 1994. The total amount of months with less precipitation than the average was approximately 9-10 months. In these years, four to six months had precipitation amounts below 50% of the average. Some locations (e.g., Lom, Sadovo, Stara Zagora, Sliven) even had seven and nine dry months.

It is necessary to remark that there were no full years with only dry (precipitation below average) or wet months (precipitation above average) months during the period with instrumental records (more than 100 years) in Bulgaria. Particular months or successive months with precipitation above the average were registered in the driest years of the 20th century.

Recent Drought Studies

Interesting aspects of the Bulgarian drought have been observed during the last decades (e.g., Alexandrov 1995; Alexandrov and Slavov 1998; Slavov and Alexandrov 1993, 1994a, 1994b, 1998; Slavov and Georgieva 1998; Tomov *et al.* 1995, 1996). Investigations have been directed to the analysis of annual and seasonal precipitation, especially precipitation during the warm (April-September) and cold (October-March) halves of the year. Precipitation sums during the potential crop-growing season (characterized by air temperature above a base of 5° or 10°C), actual growing season (from sowing to maturity) and non-growing period (characterized by air temperature below a base of 5°C) were also investigated.

Precipitation during the non-growing period is very important for soil moisture recharge. Precipitation is especially important during the crop-growing season for crop water consumption. Assessments of precipitation amounts, intensity, and distribution are necessary for proper irrigation management, developing appropriate technology for crop cultivation, and guiding agrotechnological management (e.g., sowing and harvesting). Reduced precipitation causes decreases in crop production. Precipitation is also an important factor in recharging water resources (e.g., reservoirs, lakes, ground water) used for irrigation.

Although precipitation in Bulgaria is important throughout the year, precipitation during the warm half of the year (from April to September) is especially important for agriculture. The distribution of seasonal precipitation during this period is shown in Figure 5.1. The average national precipitation during this time of the year is about 330mm. Maximum precipitation is 570mm in the Balkan Mountains, and minimum precipitation is 210mm. Precipitation in the eastern and southeastern areas and in the Struma valley is below 300mm, an amount assumed to be insufficient for normal crop growth, development and yield.

Drought, a major issue during the growing season, has a significant impact on the crop yield potential. Figure 5.2a represents the long-term variations of precipitation in the country during the potential crop-growing

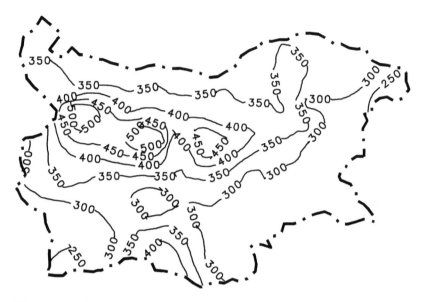

Figure 5.1   Spatial distribution of precipitation (mm) in Bulgaria during the period April-September

season, which has a base of 10°C. Significant drought conditions were observed during the potential crop-growing season in 1907, 1945, 1965 and 1985. A negative trend of precipitation during the potential crop-growing season from the end of the 1970s until the end of 1997 was found. Precipitation was below the 1961-1990 average for 16 of the last 17 years of this study. In fact, since 1982, the country has experienced more than five years of drought conditions of various intensities in various locations. The drought situation had severe impacts on agriculture in 1992, 1993 and 1996. Such a long drought during the potential crop growing season (when temperatures exceed a base of 10°C) as the most recent one had not been observed during the 20th century.

Similar drought periods during the last decade were observed during the non-growing period except in 1996 and 1997. Most of the years during the second half of the 1980s and first half of the 1990s were characterized by negative precipitation anomalies (Figure 5.2b).

Precipitation during the period from July to September characterizes summer and early autumn conditions. Figure 5.3 shows the long-term variation of precipitation during this period. Seasonal precipitation varied from year to year during the study period. In some years, very low summer precipitation ushered in droughts of different intensities. Bulgaria experienced

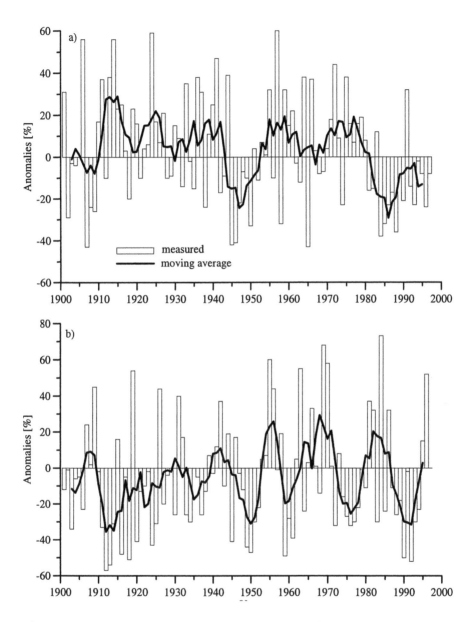

Figure 5.2 Precipitation anomalies in Bulgaria during the potential crop-growing season above a base of 10°C (a) and non-growing period below a base of 5°C (b), relative to the current climate (1961-1990)

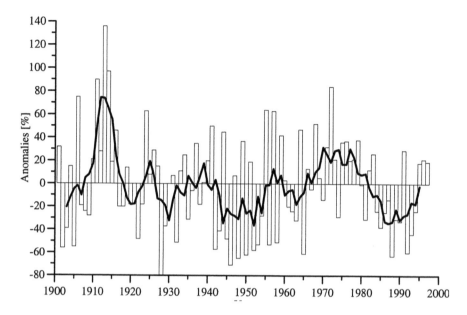

Figure 5.3 Precipitation anomalies in Bulgaria from July to September, relative to the period 1961-1990

several summer drought episodes during the 20th century, most notably in the 1940s, 1950s, 1980s and 1990s. The solid curve in Figure 5.3 averages out year-to-year fluctuations and shows the long-term variations in precipitation. This filtered curve suggests that there may have been a decreasing trend in precipitation rates during July-September since the end of the 1970s.

Looking closer at definitions based solely on rainfall, it is evident that a number of these refer to short-term "droughts" or better, "dry spells." Typical examples of these are 10 or 15 days with no rain. These appear to be tied mainly to climatic conditions endemic to Europe where rainfall normally occurs at fairly frequent intervals. Because of this, crop and animal husbandry and water storage operations are not geared for long spells of rainless weather which are seasonally normal in some semi-arid regions. The long-term fluctuations in the number of dry spells (periods of 10 to 15 days with reduced rain) during the potential crop-growing season are presented in Figure 5.4. It can be seen that there has been an increase in the number of 10-day dry spells since the beginning of the 1970s.

Normally, the actual (sowing-maturity) and potential crop-growing seasons do not coincide. The first season begins later and ends earlier than the second season. The average national precipitation during the actual growing season of maize (group 600 according to the FAO classification) ranged from

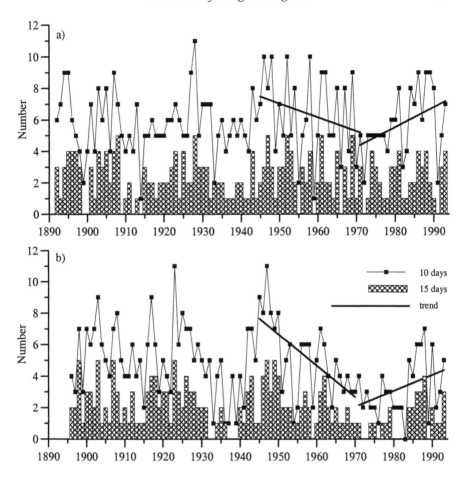

Figure 5.4   Long-term fluctuations of annual dry spells of 10 to 15 days during the potential crop-growing season above a base temperature of 10°C in Sadovo (a) and Kyustendil (b)

220-280mm during the period 1961-1997. These normal values are considered insufficient for normal crop growth. The maximum and minimum values were equal to 540mm and 110mm, respectively. Summer precipitation less than 200mm is considered insignificant for normal maize growth, development, and yield formation, especially during the last years of the 20th century. The actual growing season of maize consists of vegetative (sowing-silking) and reproductive (silking-maturation) stages. The solid line in Figure 5.5 represents trends in rainfall amounts, obtained using a linear fit method. Obviously, there have been decreasing trends in precipitation in the course of the actual and

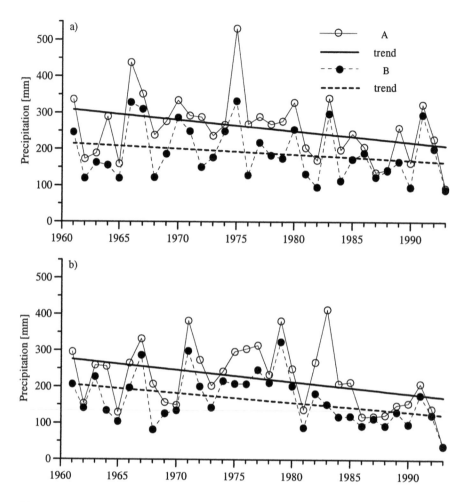

Figure 5.5 **Precipitation fluctuations and trends during the period of the actual growing season (A) and the vegetative period (B) of maize in Kneja (a) and Plovdiv (b)**

vegetative growing seasons since 1961. Precipitation in Plovdiv was considerably reduced after 1986.

Usually drought in Bulgaria occurs over most of the national territory. That is why severe droughts greatly damage agricultural crop production. Significant losses in agriculture can be observed when variations in weather conditions affect crop growth and development, reducing yields below 50% of average crop yields. The impact of weather conditions on the grain yield of the most important cereals in the country, winter wheat and maize, was also investigated during the last decades. Attempts were made to assess the degree of impact of weather conditions during different phenological stages on the process of yield formation and final grain yield (see Chapter 13, this volume).

Conclusions

1. There were three significantly long droughts during the 20th century in Bulgaria: 1902-1913; 1942-1953 and 1982-1994. Dry years made up approximately 20% of the years during the first period. The dry years during the second period increased to 40% of the total second-period years, and dry years during the period 1982-1994 made up 50% of the period; droughts covered most of the area of the country;

2. Precipitation during the potential crop-growing season (April-September) is very important for agriculture in the country. The regions which are characterized by precipitation less than 300mm during the potential crop-growing season require intense irrigation in order to sustain agricultural production;

3. The actual (sowing-maturity) and potential crop-growing seasons do not coincide. The investigations of precipitation during the actual growing season of spring crops (e.g., sunflower, maize, soybean, bean) showed that when precipitation totals are less than 200mm, irrigation is imperative;

4. The driest crop-growing seasons were observed in 1907, 1945, 1965, 1985 and 1993. Precipitation during the crop-growing season has had a negative trend since the 1970s. Droughts were very frequent during the 1990s. The last drought period was the driest recorded during the 20th century.

References

Alexandrov, V. 1995. "Climate Variability and Drought in Bulgaria," pp. 35-42 in N. Tsiourtis (editor), *Water Resources Management under Drought or Water Shortage Conditions*. Rotterdam: Balkema.

Alexandrov, V. and N. Slavov. 1998. Колебания на добива на царевица в зависимост от метеорологичните условия, *Растениевъдни науки* 35(1):11-17.

Dimitrov, L. 1941. Валежни и сухи периоди в България, *Известия на метеоролгичната служба във войската*, кн. 2:149-178.

Ganev, G. and L. Krastanov. 1949. Върху колебанията на многогодишния ход на някои метеорологични елементи в България, *Год. на Соф. Университет*, Природомат. Ф-т, 49(1):468-491.

Ganev, G. and L. Krastanov. 1951. Принос към метеорологичните изследвания на засушаванията в България, *Изв. БАН*, сер. Физическа, 2:256-279.

Hershkovich, E. 1968. Повторяемост и обхват на различните по сезони и вреда на селското стопанство засушавания в България, *Сб. Характер на зас. и пром. поливен режим на селскостопански култури в България*, стр. 111-152.

Kirov, K. 1934. Пролетното засушаване и посевите през 1934, *Земеделско метеорологичен бюлетин*. ЦИМ 5:8-10.

Kirov, K. 1948. Климатологически и агроклиматологически предпоставки за изследване сушите в България, *Сб. Сушата и борбата с нея*. БАН стр. 1-71.

Kiryakov, K. 1941. Климатът на пшеницата в България, *Тр. Централният метеорологичен институт* 1:143-272.

Kiryakov, K. 1945. Сушата през 1945 година, *Сп. Земеделие*. 7-8:122-128.

Koleva, E. 1981. Многогодишни колебания на валежите и температурите на въздуха в България, *Сп. Хидрология и метеорология* 2:69-76.

Koleva, E. 1987. Многогодишни колебания в хода на температурата на въздуха и валежите в България, *Сп. Проблеми на мет. и хидр.* 2:27-40.

Koleva, E. 1991. Колебания на климата в България, Монография: *Климатът на България*, стр. 467-477.

Koleva, E. 1994. "Variability of Precipitation in the Danube Basin," pp. 425-428 in *XVIIth Conference of the Danube Countries on Hydrological Forecasting and Hydrological Based Water Management*, Hungary.

Sabeva, M. 1968. Климатична характеристика на засушаванията в България, *Сб. Характер на засуш. и пром. поливен режим на сел. култури в България*. Стр. 13-50.

Slavov, N. and K. Georgieva. 1998. Зимните и летните валежи в България през изминалото столетие и връзката им със слънчевата активност, *Сб. Пета конференция "Основни проблеми на слънчево-земните въздействия."*

Slavov, N. and V. Alexandrov. 1998. "Spring Crops in Bulgaria Damaged by 1996 Summer Drought," *Drought Network News* 10(1):4-5.

Slavov, N. and V. Alexandrov. 1989. Многогодишните колебания на валежите през неактивния и активния потенциален вегетационен период, *Сб. Селскостопанска метеорология* 3:136-141.

Slavov, N. and V. Alexandrov. 1993. "Drought in Bulgaria, Update and Historical Perspective," *Drought Network News* 5(2):12-15.

Slavov, N. and V. Alexandrov. 1994a. "Drought and Its Impact on Grain Yield of Maize," *17th European Regional Conference on Irrigation and Drainage* 1:321-327.

Slavov, N. and V. Alexandrov. 1994b. "Persistent Drought in 1993 Affects Bulgarian Agriculture," *Drought Network News* 6(1):19-20.

Slavov, N. *et al.* 2003. Климатични особености на засушаването, *Засушаването в България*. София: Българска академия на науките, стр. 38-49.

Stefanov, S. 1968. Циркулационни особености на засушаванията в България, *Сб. Характер на засушаванията и пром. поливен режим на селскост. култури в България*, стр. 51-109.

Tomov, N., N. Slavov and A. Gancheva. 1995. "Засушаването и добива от царевицата през 1993 г.," *Сп. Растениевъдни науки*, 9-10:47-52.

Tomov, N., N. Slavov and V. Alexandrov. 1996. "Drought and Maize Production in Bulgaria," pp. 169-176 in *International Symposium on Drought and Plant Production*. Beograd, Yugoslavia.

Chapter 6

Drought During the 20th Century

Ekaterina Koleva, Nicola Slavov, Vesselin Alexandrov

Introduction

At any given time, at least one nation in the world is being adversely affected by drought. Although drought is a natural component of the climate in arid and semi-arid areas, it can occur in areas which normally receive adequate precipitation. Available hydrometeorological data indicate that droughts have occurred throughout the last century in Bulgaria and that they are a natural part of the climatic cycle of the entire Balkan Peninsula.

Annual precipitation in Bulgaria ranges from 550-600mm in the lowest elevations in the country to 1000-1100mm in the highest elevations (Koleva 1988, 1989). The precipitation distribution in the country is mainly caused by synoptic atmospheric conditions over Bulgaria, which are influenced considerably by topography. Insufficient precipitation is climatically common in some parts of Bulgaria, leading some scientists to speculate that the country has a tendency toward drought. According to the Budyko drought coefficient (**K**), Bulgaria is characterized by insufficient moisture. This coefficient is calculated using data from the annual radiation balance and total annual precipitation (*Klimatalogiya* 1989). The Budyko drought coefficient is between 1.5 and 1.8 for Bulgaria.

A Century of Droughts

Precipitation distribution is one of the basic identifiers of drought occurrences in a given region. However, the distribution of additional meteorological elements should also be taken into account in order to describe the degree of drought. For example, the distribution of air temperature is an especially important characteristic for drought classifications. Usually, average precipitation for a given region is calculated from long-term variations in precipitation data. Accordingly, the index of anomaly **V** is calculated as (Koleva 1988):

$$V_j = \frac{1}{n} \sum_{i=1}^{n} \frac{x_i}{\overline{x_i}} \tag{1}$$

where: $j = 1,...,n$ years, x_i - total annual precipitation in the ith station, $\overline{x_i}$ - averaged annual precipitation for the same station, n – station number.

Figure 6.1 shows the long-term variations of annual and seasonal precipitation in different regions for the period May-September. There is a clearly decreasing trend in precipitation at the end of the 20th century. The last 20 years were drier and warmer than normal during the so-called "current" climatic period of 1961-1990 (Koleva *et al.* 1996). Annual precipitation was approximately 80-85% of normal, and the winters were especially dry. Air temperature in January was higher than the current climatic average. July precipitation was close to normal, but air temperature was above normal. Similar drought conditions during the 20th century were also observed from 1945 to 1953 (Koleva 1995). Figure 6.2 presents annual precipitation variability for the mountains. A significant decrease in recent years is easily seen.

Statistical methods were used to analyze long-term variations in precipitation, air temperature, wind speed, and direction. The long-term data series were smoothed by averages and approximated by means of a polynomial. These methods eliminate the random and short periodical fluctuations of the time series. Because visual evaluation of the smoothed time series is a subjective method, the Spearman coefficient (r) and the Mann-Kendall coefficient (r_1) were used to investigate the existence of lasting trends (WMO 1966, 1990).

The statistical characteristics for the Danube Plain, Thracian Lowland, and various mountain stations in the country are presented in Tables 6.1 and 6.2. The precipitation trend is negative for the period 1931-1996 and in some places (marked with *) it is statistically significant at the 5% probability level (Koleva and Iotova 1992).

Different quantitative criteria were also used for a comparison of the drought frequency and intensity between different regions and years. For these comparisons, two drought indexes, Ped and de Martonne, were used (Koleva 1998; de Martonne 1925; Ped 1975):

Ped index: $$P_{ed} = \frac{\Delta T}{\sigma_T} - \frac{\Delta P}{\sigma_P} \tag{2}$$

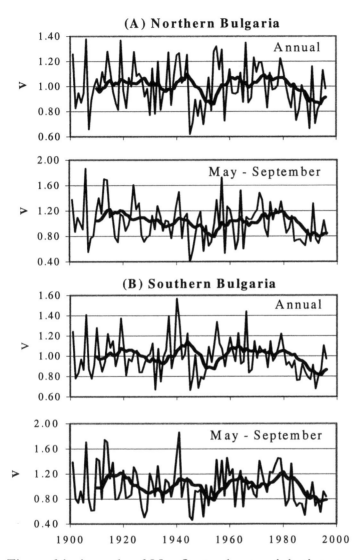

Figure 6.1 Annual and May-September precipitation anomaly index (V) and 10-year moving averages for (A) northern Bulgaria and (B) southern Bulgaria, 1901-1996

where: ΔT and ΔP – anomalies of air temperature and precipitation, relative to a given time period; σ_T, σ_P – standard deviations of air temperature and precipitation. Values of the Ped index that are between 1 and 2 indicate the existence of an insignificant drought, values of $2 <$ Ped <3 indicate moderate

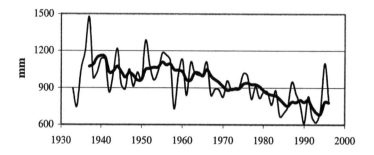

Figure 6.2 Variation of annual precipitation and 10-year moving average for mountainous areas

drought, and values Ped > 3 indicate severe drought. Negative index values characterize a wet period. Droughts are more frequently observed in the Thracian Lowland, where the Ped values are higher than 2. Drought years during the last decade of the 20th century were 1989, 1990, 1992 and 1994. The last three years were especially dry, when the Ped values were higher than 3 and even 4. In general, there is an increasing trend of the Ped index at the end of the century (Figure 6.3).

$$\textbf{De Martonne index:} \qquad J = \frac{\dfrac{P}{T+10} + \dfrac{12p}{t+10}}{2} \tag{3}$$

where: **P** and **T** equal total annual precipitation and annual air temperature, **p** – precipitation during the driest month of the year, **t** – air temperature during the warmest month of the year. When the **J** index is less than 30, drought conditions may be observed. A value of less than 20 is typical of severe drought.

Table 6.1 Mann-Kendall precipitation characteristics

Seasons	Thracian Lowland	Danube Plain
Winter	-0.26*	-0.11
Spring	-0.20*	-0.11
Summer	-0.19*	-0.24*
Autumn	-0.32*	-0.16
May-September	-0.17	-0.15
Annual	-0.50*	-0.30*

* 5% significance level

Table 6.2 Rank statistics of annual precipitation in the mountains

Station	Spearman (r)	Mann-Kendall (r_1)
Borovets	-4.06*	-0.34*
Sitnyakovo	-1.65	-0.02
Beli Iskar	-6.80*	-0.49*
Musala Peak	-8.54*	-0.51*
Stoletov Peak	-1.60	-0.11
Ambaritsa	-2.17*	-0.20*
Botev Peak	-0.56	-0.09
Boeritsa	-1.29	-0.10
Cherni Vruh	-4.42*	-0.37*

* 5% significance level

The average value of the J index was about 25 in different regions during the period 1961-1990. It decreased to 19-23 during the period 1982-1994. Figure 6.4 shows the variations of the J index for every month of the year for these two periods. July, August and September are the driest months with index values of about ten.

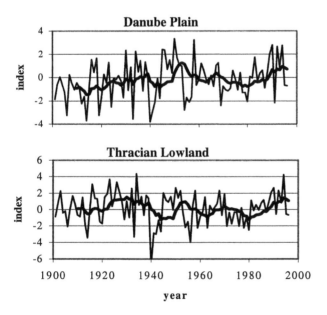

Figure 6.3 Annual Ped index and 10-year moving average for the Danube Plain and Thracian Lowland

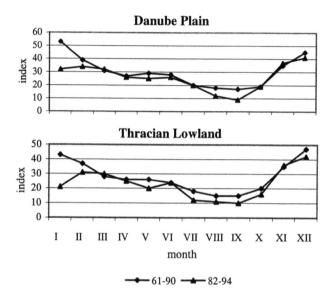

**Figure 6.4 Monthly de Martonne Index for the Danube
Plain and Thracian Lowland**

This analysis shows that the study period can be divided into sub-periods
with duration of 10-15 years. These years are characterized by different
precipitation conditions. The following criteria were used in order to
determine the moisture conditions:

$$P < \overline{P} + 2\sigma_p \quad \text{- severe drought} \tag{4}$$

$$\overline{P} + 2\sigma_p < P < \overline{P} + \sigma_p \quad \text{- drought} \tag{5}$$

$$\overline{P} - \sigma_p < P < \overline{P} + \sigma_p \quad \text{- normal} \tag{6}$$

$$P > \overline{P} + \sigma_p \quad \text{- wet} \tag{7}$$

where: P – precipitation in a particular year, $\overline{P}$ – average precipitation during
the period 1961-1990, σ_p – standard deviation. The distribution of the years
according to these criteria is presented in Figure 6.5.

Three chief periods during the 20th century are characterized by long and
severe droughts, namely 1902-1913, 1942-1953 and 1982-1994. During the
first period, the drought years were approximately 20% of the total years.
However, they increased to 40% of the second period and even approximated

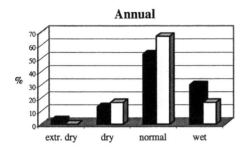

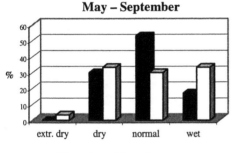

■ northern □ southern

Figure 6.5   Annual and May-September distribution (%) of extremely dry, dry, normal, and wet years for the period 1961-1990

50% of the last period 1982-1994. Another characteristic of the last period is that years with above normal precipitation were not observed in southern Bulgaria (Figures 6.6 and 6.7). Average precipitation in the Danube Plain was about 560mm and it was 540mm in the Thracian Lowland. Precipitation in the Danube Plain and Thracian Lowland during the period May-September was 280 and 230mm, respectively. The coefficient of variation (C_v) in the two regions was about 0.25-0.35, which signifies insignificant moisture availability or a dry climate.

The driest year during the 20th century study period (1901-1996) was 1945. Other considerable dry years were 1902, 1907, 1932, 1934, 1946, 1948, 1950, 1953, 1985, 1986, 1990, 1992 and 1993. It is important to note that there are some differences in the classification of dry years between the regions of northern and southern Bulgaria.

Figure 6.8 shows the monthly distribution of precipitation during the periods 1961-1990, 1942-1953 and 1982-1994. It can be seen that during the period 1942-1953 drought conditions were observed at the end of summer and at the beginning of autumn when precipitation was 10-20mm less than

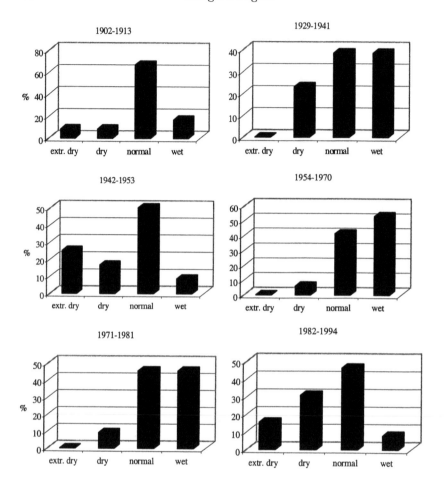

Figure 6.6　Distribution (%) of extremely dry, dry, normal, and wet years for northern Bulgaria for different periods of the 20th century

normal. Precipitation amounts were especially low during the winters of the period 1982-1994. The characteristics of the precipitation distribution during the aforementioned time periods are similar to those of a moderate continental climate.

　　Particular years, months or successive months with insignificant precipitation, i.e. periods of atmospheric drought, are not uncommon events in the country and can even be considered as normal climatic characteristics of Bulgaria. Climatically, about two successive dry months occur in the lower areas of the country annually. However, there are some years with six and

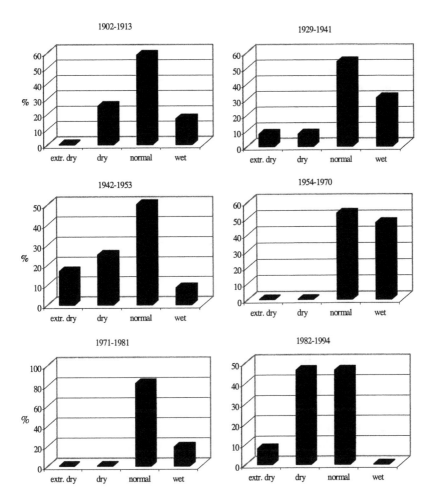

Figure 6.7 Distribution (%) of extremely dry, dry, normal, and wet years for southern Bulgaria for different periods of the 20ᵗʰ century

more successive dry months. Long dry periods during the cold-half of the year were observed in 1913, 1934, 1967, 1976 and 1983. Extended dry periods during the warm-half of the year occurred in 1928, 1945, 1965 and 1985.

Climatically, there are three successive months during the year with below normal precipitation. The number of months with below normal precipitation throughout the year is about five. Usually, precipitation is less than 50% of the average in two-three months during the year. In some years, 5-6 or even more successive months with less precipitation than the average can be observed. This occurred about 25-30% of the years during the period

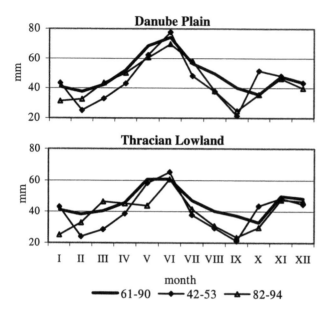

Figure 6.8 Monthly precipitation distribution in the Danube Plain and Thracian Lowland for the current climate (1961-1990) and the two driest periods of the 20th century

1906-1996. In 1945, precipitation amounts were especially low from January to November in most regions of the country. The 1944 autumn was also dry, making the total drought duration approximately 12-15 months. Several successive dry months (from January to August and even until October in some places) were also observed in 1938, 1959 and 1968. Precipitation in some months was 20-30% lower than the norm. These drought episodes, however, were not observed in all regions of Bulgaria. There were from four to six successive months with reduced precipitation in 1985, 1990, 1992, 1993 and 1994. The total months with lower precipitation than the average in some weather stations were about nine or ten. In these years precipitation was 50% less than the average for four, five, or sometimes six months. Even in some weather stations such as Lom, Stara Zagora and Sliven dry months lasted for seven to nine months.

It is necessary to emphasize that during the 20th century there was no year with only above-average monthly precipitation. In Bulgaria, there was always at least one month or successive wet months during the most severe drought years.

The 1982-1994 Drought

Meteorological features of the last long drought period during the 20th century (1982-1994) are described in the second part of this chapter. Results reported by Sharov *et al.* (1994) were used.

Monthly precipitation during the period May to November in 1982 was less than the average. Precipitation from January to May in 1983 was also below average and weather was warmer than usual. Significant positive anomalies of air temperature (2-4°C) were measured in April. The summer was characterized by lower air temperatures and increased precipitation. Precipitation was near average during the second half of the summer, although there were significant precipitation variations in the different weather stations. Air temperature in November was lower than normal, however, precipitation amounts were low.

The winter of 1984 was relatively warm (especially in January) with increased precipitation. The winter extended into March when snowfall was a frequent event and air temperature was lower than normal. Monthly precipitation was somewhat reduced from April until the end of the year. Only precipitation in August was four to five times above the average in southeastern Bulgaria. Maximum precipitation was measured on August 20.

The winter of 1985 was cold, especially in northern Bulgaria. Precipitation was near average. Snow cover was deep and frozen. Monthly precipitation from March to October was less than average. A temporary increase in precipitation was measured in August in northwestern Bulgaria and in September in northeastern Bulgaria. Precipitation was above average in November, however, precipitation dropped to below normal in December. The months with air temperature higher than the average (e.g., April, May, August, December) predominated in 1985. Air temperatures lower than average were measured during the winter and in the month of October.

Precipitation from March until the end of 1986 was below the average. An increase in precipitation to above average amounts was observed only in October in southeastern Bulgaria and in November in western and central Bulgaria.

A significant precipitation deficit was observed from July to September in 1992. This drought period continued into the autumn and also during the winter season in 1993. Average precipitation in the country was 60%, 80% and 5% of the average for the last three months, respectively, in 1992. Precipitation significantly decreased in June 1993. Summer precipitation amounts were unusually low. Precipitation in western Bulgaria was trivial from the beginning of June until the end of August–between 25 and 40% of the average. The precipitation amounts in eastern Bulgaria were relatively higher than in western Bulgaria; however, they were still less than the average. Actually, summer precipitation accumulated mainly because of the rainfall during the first half of June and at the end of August. The rest of this summer period was characterized by short rainfall with intermittent dry spells. It was a severe, long drought, which began in the middle of 1992. The intensity of this drought, especially in western Bulgaria, is comparable only to the droughts of 1928 and 1958.

The drought in 1993 in Bulgaria was strongly related to atmospheric circulation. The atmospheric circulation over southern Europe and the Mediterranean was not typical for the winter and spring seasons. Almost no Mediterranean cyclones developed, and regional weather was influenced by a stable anticyclone with high atmospheric pressure. During the beginning of the summer, high atmospheric pressure was also recorded over Western Europe. The Azores anticyclone continued to impact the Balkan Peninsula and the air masses over Bulgaria did not produce significant precipitation. The cyclone center over the Scandinavian Peninsula additionally influenced drought conditions in the country. Atmospheric fronts were moving rapidly and frequently over northern Bulgaria. Strong wind was blowing toward the Black Sea coast, which further decreased air humidity. The atmospheric circulation over eastern Europe in the middle of the summer created anticyclones with tropical air masses causing warmer weather. Air temperatures in Bulgaria under these conditions increased to 34-39°C.

Consistent with the atmospheric circulation over Bulgaria, summer weather anomalies, even more significant than those in 1992, were also observed during some years of the 20th century. For example, unusually dry and hot summers were observed in 1928, 1946 and 1958.

The last long drought period ended in 1994. That year was also warm and dry. The average air temperature during the first eight months of the year was higher than average. The air temperature anomalies in some locations in northern Bulgaria were even above 2°C. Precipitation was below climatic norms. Less precipitation was measured in central and southeastern Bulgaria. The amounts were from 50 to 70% less than the average. The winter was unusually warm and dry. The negative precipitation anomalies decreased during the spring season. However, the tendency for low precipitation amounts was still stable. The weather conditions improved in June, although precipitation was still below the average. In general, agriculture in the country, especially crop growing, was negatively affected by drought conditions in 1994.

It is interesting to analyze the basic features of atmospheric circulation that caused the weather anomalies in 1994. During the last years, a decreasing trend of Mediterranean cyclones passing over Bulgaria has been observed in winter and spring. That is because the Azores anticyclone was well developed and often influenced the southern regions of the European continent. Anticyclones and warm air masses influence weather in Bulgaria. The development of the Azores anticyclone in spring and during the first half of the summer was oriented toward the Scandinavian Peninsula. That was the reason for a decrease in intensity of the west-east air mass movement in Europe. The anticyclone effect increased in Central Europe.

As a result, the atmospheric circulation over the Balkan Peninsula changed in the summer of 1994, mitigating the drought. One of the effects of this kind of atmospheric system is an increase in the number of polar air masses over Eastern Europe. Although these atmospheric processes do not lead directly to an increase in precipitation in the region, they can be an indication of a decrease in existing dry, warm weather which started in the beginning of the 1990s. Winter precipitation in 1995 and 1996 was significant, with amounts 20% above average. 1997 was also characterized by wet conditions. However, it is necessary to keep in mind that dry periods may occur after every wet period. Proof of this statement is the year 2000. The last year of the 20th century was among the driest years since instrumental meteorological records became available in Bulgaria.

References

De Martonne, E. 1925. *Traite de Geographie Physique*. Paris: Colin.

Klimatalogiya. 1989. *Климатология*. Ленинград: Гидрометеоиздат.

Koleva, E. 1981 Многогодишни колебания на температурата и валежите в България, *Хидрология и Метеорология*, кн. 2.

Koleva, E. 1987. Многогодишни колебания в хода на температурата и валежите, *Проблеми на метеор. и хидролог.*, кн. 2.

Koleva, E. 1988. Особености в разпределението на валежите в равнинната част на България, *Проблеми на метеорологията и хидрологията*, 2:41-48.

Koleva, E. 1991. Разпределение на валежите, *Климатът на България*. Изд. БАН, София, 499.

Koleva, E. 1995. "Drought in the Lower Danube Basin," *Drought Network News* 7(1):6-7.

Koleva, E. and A. Iotova. 1992. "Precipitation Variability in Bulgarian Mountains," *Proceedings of the 22nd International Conference on Alpine Meteorology*, Toulouse, France, pp. 421-424.

Koleva, E., C. Boroneant, and E. Bruci. 1998. "Study on the Variability of Annual and Seasonal Precipitation over the Balkan Peninsula," vol. 3, pp. 1084-1089 in *The Second International Conference on the Climate and Water*, Esoo, Finland, 17-20 August,

Koleva, E., L. Krastev, E. Peneva, E. Stanev. 1996. "Verification of High Resolution Climatic Simulations for the Area of Bulgaria, Part I: The State of the Climate for the Period 1961-1990," *Bulgarian Journal of Meteorology & Hydrology*, 7(3-4):73-84.

Ped, D. A. 1975. О показателе засухи и избыточного увлажнения. *Труды Хидрометцентра CCCP* 156:19-38

Sharov, V., N. Slavov, E. Koleva, P. Ivanov, E. Moraliiskı, L. Latınov, S. Dakova, A. Gocheva, R. Peneva, V. Alexandrov. 1994. Засушаванията в България, *Отчет по договор НИ-НЗ-19/1991* от НФ "Научни изследвания" към МОН, стр. 65.

WMO. 1966. *Climate Change.* World Meteorological Organization Technical Note 9, Geneva.

WMO. 1990. *On the Statistical Analysis of Series of Observations.* World Meteorological Organization Technical Note 143, Geneva.

Chapter 7

The Balkan and European Context
of the Drought

Liem Tran, C. Gregory Knight, Victoria Wesner

It is helpful to understand the drought period in Bulgaria in the context of climatic patterns in the Balkans and Europe, since the major determinants of climate in Bulgaria include both hemispheric and local factors. At the hemispheric level, climate in Bulgaria and the Balkans is strongly influenced by the Icelandic low (cyclone) and the Azores high (anticyclone) over the Atlantic Ocean, as well as by seasonal pressure features, including the winter Mediterranean low and European and Siberian highs. At a more local level, climate is strongly influenced by topography in which higher elevation leads to decreased temperature and evaporation and increased precipitation, in addition to which mountain areas act as topographic barriers to circulation in the lower atmosphere. Thus Bulgaria's climate is determined at different scales in geographic space as well as by changing temporal patterns of atmospheric activity.

In this chapter, we use larger-scale data sets that incorporate Bulgaria, the Balkans, and Europe. Several research organizations have developed grid-based climate data sets that are useful for seeing Bulgarian climate in larger contexts. One of these data sets is the 0.5° gridded monthly precipitation data from the Climate Research Unit (CRU) at the University of East Anglia for the period 1901-1995. This data set has a cell resolution of about 55km (north-south) by 40km (east-west) over Bulgaria. Such a grid size still hides considerable topographic variability. However, Cavazos (2000) found a reasonable agreement between CRU data and selected station data in Bulgaria. In another test, we found excellent agreement between observed data at Sofia (42.39N, 23.23E) (NOAA, National Climatic Data Center) and CRU grid-point data at (42.75, 23.25) and (42.25, 23.25) for the period 1987-1995. The correlation coefficients were 0.98 and 0.96, respectively. For comparative purposes, we use 1961-1990 30-year averages for each grid cell, the standard period for climatological norms.

For analysis of atmospheric dynamics over Europe during 1948-2000, we used monthly gridded 2.5 degree resolution data from the Global Reanalysis

Project of the U. S. National Centers for Environmental Protection-National Center for Climate Research (Kalnay 1996). This data set provides a unified global picture of atmospheric dynamics, including sea level pressure (SLP) and geopotential heights at the 500-hPa level (GPH-500), both of which allow visualization of lower and upper atmospheric pressure fields which drive (or block) moisture-bearing air masses that bring precipitation.

In another paper, we describe the development of a method based on self-organizing maps to conceptualize atmospheric patterns that dominate seasonal climate conditions influencing Bulgaria (Tran *et al.* 2004). This technique was also used by Cavazos (2000) to explore relationships between atmospheric synoptic conditions and winter precipitation in Bulgaria. An additional paper (Tran *et al.* 2002) developed a spatial-dryness index for analysis of drought, incorporating both severity of low precipitation and its geographic extent. Furthermore, we applied the spatial-dryness index and self-organizing maps to link drought patterns to synoptic conditions. In this chapter we draw upon these techniques to help understand the recent drought occurrences in Bulgaria.

Drought in Bulgaria

There have been many studies of drought in Bulgaria, including chapters in this book. Many of these studies point to a continued drying of the Balkan-Bulgarian region (Sharov *et al.* 1994; Velev 1996; Sahsamanoglou *et al.* 1997; Schonweise and Rapp 1997; Tran *et al.* 2002 Koleva *et al.*, this volume). The spatial-dryness (SD) index for Bulgaria is defined as the proportion of the country that has precipitation in the lowest quintile of record (lowest 20 percent of recorded rainfall for the respective time period, or alternatively, the 80 percent confidence level). From the CRU data from 1901-1995, we developed an annual SD index for Bulgaria (using 84 grid cells covering the area from 41.25N to 44.25N lat. and from 22.25E to 27.75E long.), finding that there were 17 occurrences of annual SD index greater than 50 percent during that time. The period 1901-1903 had three consecutive years of SD index 73 or greater; 1985 and 1986 had SD = 73.8 and 67.9, respectively; and for 1992 and 1993, SD = 73.8 and 67.9 (Tran *et al.* 2002). Paradoxically, years 1994 and early 1995 when the greatest drought impacts were being experienced, the national annual SD index showed only SD = 5.2 and 0.0, respectively. The year 1994 was relatively dry (much of the nation in the second quintile of precipitation) but not an extremely dry year compared with others in the period 1982-1995 (Table 7.1). However, early 1994 (January to

Table 7.1 SD index by year for the period 1982-1995. (I is the driest, V is the wettest; values indicate proportion of Bulgaria receiving respective quintiles of precipitation)

| Year | Quintiles | | | | | SD |
	I	II	III	IV	V	
1982	48.8	26.2	15.5	6.0	3.6	**48.8**
1983	20.2	29.8	22.6	25.0	2.4	**20.2**
1984	0.0	0.0	11.9	65.5	22.6	**0.0**
1985	73.8	17.9	8.3	0.0	0.0	**73.8**
1986	67.9	29.8	2.4	0.0	0.0	**67.9**
1987	2.4	11.9	32.1	45.2	8.3	**2.4**
1988	19.0	34.5	35.7	10.7	0.0	**19.0**
1989	36.9	56.0	7.1	0.0	0.0	**36.9**
1990	61.9	38.1	0.0	0.0	0.0	**61.9**
1991	1.2	16.7	13.1	20.2	48.8	**1.2**
1992	84.5	15.5	0.0	0.0	0.0	**84.5**
1993	67.9	21.4	8.3	2.4	0.0	**67.9**
1994	4.8	53.6	33.3	8.3	0.0	**4.8**
1995	0.0	0.0	3.6	16.7	79.8	**0.0**

March) continued the 1993 drought, giving evidence that after a severe drought (e.g., the 1992-1993 event), several more consecutive months with dry conditions will continue to cause significant negative impacts in Bulgaria.

With slightly less than a century of data, there were three occurrences of two or three consecutive years in which two-thirds of Bulgaria (or more) received rainfall in the lowest 20 percent of recorded annual precipitation. In addition, at least one year in six has half of the country receiving less than this amount of precipitation. Clearly, understanding and planning for drought is imperative, particularly if Bulgaria becomes warmer and relatively drier in future as a result of global climate change.

The 1982-1994 Dry Period

Several documents have discussed the dry period of 1982-1994 and the drought in 1992-1993 (for example, Sharov *et al.* 1994a, 1994b; Knight *et al.* 1995; Velev 1996). Here we provide a closer look at this dry period through visualization. Annual precipitation from 1981 to 1995 and annual wet/dry conditions in comparison with the 1961-1990 average are shown in Figure 7.1. The map series reveals extremely dry conditions for different parts of the country in different years. For example, the northeast area was hit hard in

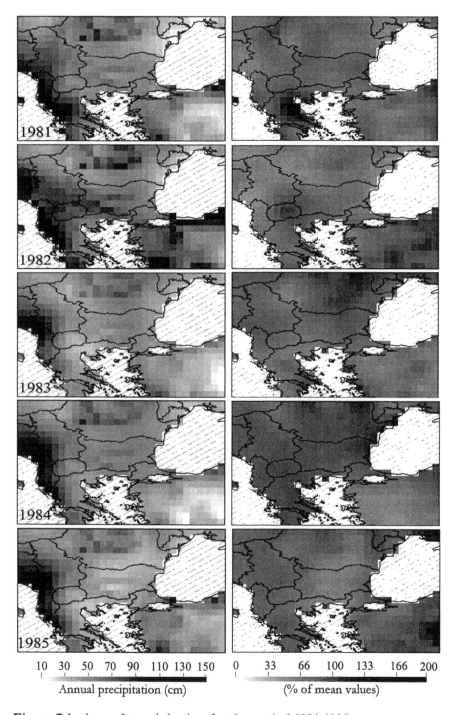

Figure 7.1 Annual precipitation for the period 1981-1995

Figure 7.1 Annual precipitation for the period 1981-1995 (continued)

Figure 7.1 Annual precipitation for the period 1981-1995 (continued)

1992 but less severely in 1993 while the southwest was extremely dry in 1993. Table 7.1 indicates the SD index for Bulgaria during the same period. The period from 1982-1994 had five of the twenty driest years (of the period 1901-1995) as indicated by the SD index (Tran *et al.* 2002). In addition, it should be noted that the case of two consecutive extremely dry years happened twice in this period (1985-1986 and 1992-1993). The impact of such a prolonged dry period was profound, culminating with the 1992-1993 drought.

The 1993 Drought

Figure 7.2 provides a closer look at the monthly precipitation of the Balkan region in 1993 and the monthly relative dry/wet conditions in comparison to the 1961-1990 average. Actually the 1993 drought started in December 1992, dry conditions occupied the whole country, persisting through the winter of 1993 until April (with small break in March 1993). Since then dry conditions persisted through the winter of 1994 with the exception of May and November of 1993.

Figures 7.3-7.6 provide the monthly GPH-500 and SLP patterns to help visualize changes in the atmospheric circulation over the country during the 1993 drought. In this period, anticyclonic conditions and relative high-pressure periods without surface pressure field gradients across the country were abnormally dominant, in agreement with analyses by Knight *et al.* (1995) and Velev (1996). More than usual blocking anticyclones occurred over western and eastern Europe during which Bulgaria happened to be either in their eastern (cold) or western (warm) periphery. The Atlantic cyclones in such conditions passed over the Scandinavian region and the Baltic Sea to the north of Bulgaria. Furthermore, the Mediterranean cyclones, important for generating winter precipitation, were much fewer than normal in that period. This can be seen by comparing the 1993 monthly synoptic conditions, provided in Figures 7.3 and 7.5, and the mean monthly synoptic conditions, displayed in Figures 7.4 and 7.6 (Figures 7.3-7.6 follow the text).

Drought Persistence

In other work (Tran *et al.* 2002), we examined month-to-month and year-to-year persistence of dry and wet conditions. Again using the SD index, we found that there is a greater number of dry months (7) with high probability

Drought in Bulgaria

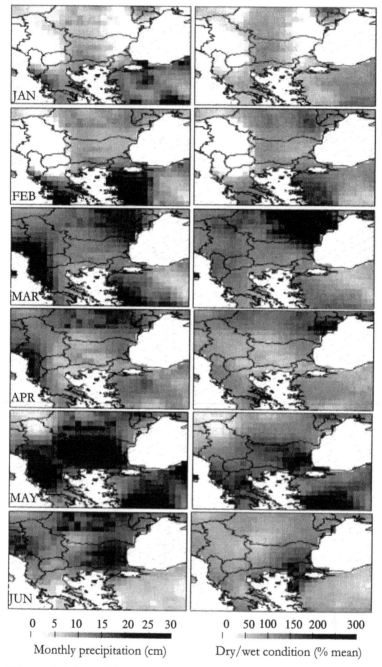

0   5   10   15   20 25   30   0   50 100 150 200   300

Monthly precipitation (cm)   Dry/wet condition (% mean)

Figure 7.2 Monthly precipitation, January-December 1993

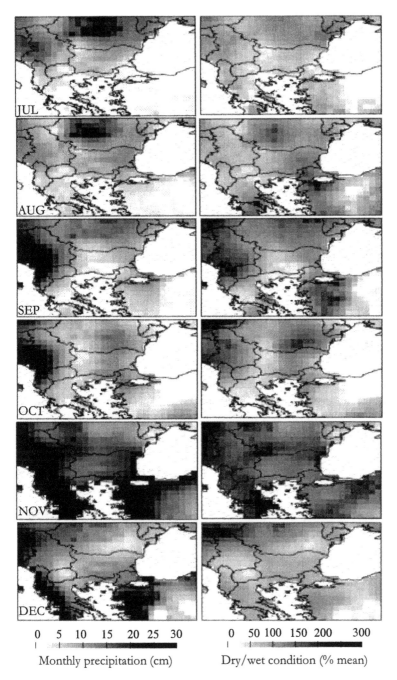

Figure 7.2   Monthly precipitation, January-December 1993 (continued)

of being followed by a wet month (5), of which only two wet transitions are significantly more likely than dry transitions. On an annual basis, we also found that both dry and wet years were more likely to be followed by a like year at statistically significant levels. We believe the phenomena of persistence is related to the occurrence of certain seasonal synoptic types as revealed by the self-organizing map analysis.

In addition, using the CRU data, we identified the frequency with which each grid cell had precipitation less than 20% of the 1961-1990 averages (Figure 7.7, following the text). We found that monthly drought frequency was greatest in winter (January through March) and late summer through early autumn (August through October). The occurrence of winter drought is particularly important, since winter snow accumulation is a critical source of spring-summer runoff in Bulgarian lowland streams.

The Future

It is impossible to determine with certainty whether drought will become more frequent in the future. Certainly, general precipitation trends are negative, and recent decades have brought a greater occurrence of extremely serious drought as measured by the spatial dryness index (SD). Continental scale analyses of European water resources by Arnell (1999), using the same spatial resolution as the CRU data set, suggests Bulgaria will have significant negative changes in runoff as a result of global climate change; other work by Chang *et al.* (2002) for southwestern Bulgaria suggests less annual change in runoff but a significant seasonal shift to winter runoff with decreased snow accumulation in mountain areas and marked decrease in summer runoff. Taken as a whole, precipitation trends, increasing occurrence of severe drought, and runoff projections based on climate change scenarios all point to seasonal and/or annual drought as a severe threat to Bulgaria. Planners and decision-makers in climate-sensitive sectors should take heed and not rely solely on past climate norms as indicative of the future.

References

Arnell, N. 1999. "The Effect of Climate Change on Hydrological Regimes in Europe: A Continental Perspective," *Global Environmental Change* 9:5-23.

Cavazos, T. 2000. "Using Self-Organizing Maps to Investigate Extreme Climatic Events: An Application to Wintertime Precipitation in the Balkans," *Journal of Climate* 13:1718-1732.

Chang, H., C. G. Knight, and M. P. Staneva. 2002. "Water Resource Impacts of Climate Change in Southwestern Bulgaria," *GeoJournal* 57:159-168.

Kalnay, E. 1996. "The NCEP/NCAR 40-Year Reanalysis Project," *Bulletin of the American Meteorological Society* 77:437-471.

Knight, C. G., S. Velev and M. P. Staneva. 1995. "The Emerging Water Crisis in Bulgaria," *GeoJournal* 35:415-423.

Koleva, E. 1988. Особености в разпределението на валежите в равнинната част на България, *Проблеми на метеорологията и хидрологията*, Кн. 5, БАН, София.

NOAA. n.d. *The NCEP/NCAR Reanalysis Project.* National Oceanic and Atmospheric Administration. http://www.cdc.noaa.gov/reanalysis

Sahsamanoglou, H., T. Makrogiannis, N. Hatzianastasiou, N. Rammos. 1997. "Long Term Change of Precipitation over the Balkan Peninsula," pp. 111-124 in Ghazi *et al.* (editors), *Eastern Europe And Global Climate Change*, European Commision.

Schonwiese, C. D. and J. Rapp. 1997. *Climate Trend Atlas of Europe.* Boston, MA: Kluwer Academic Publishers.

Sharov, V., P. Ivanov and V. Alexandrov. 1994a. "Meteorological and Agrometeorological Aspects of Drought in Bulgaria," *Romanian Journal of Hydrology & Water Resources* 1(2):163-172.

Sharov, V. *et al.* 1994b. *Drought in Bulgaria, Final Report on Contract No.NI-NZ-19/1991,* Sofia, 80 pp. (in Bulgarian).

Tran, L. T., C. G. Knight and V. Wesner. 2002. "Drought in Bulgaria and Atmospheric Synoptic Conditions over Europe," *GeoJournal* 57:165-173.

Tran, L. T., C. G. Knight, V. Wesner, A. Dean. 2004. "Connections between Monthly Atmospheric Conditions over Europe and Precipitation in Bulgaria," *In Review.*

Velev, S. 1996. "Is Bulgaria Becoming Warmer and Drier?" *GeoJournal* 40:363-370.

Acknowledgements

This work was supported by a grant to the Pennsylvania State University from the U. S. National Science Foundation Program on Human Dimensions of Global Change (SBR-9521952), Center for Integrated Regional Assessment. Views expressed are those of the authors and not those of our sponsor.

Drought in Bulgaria

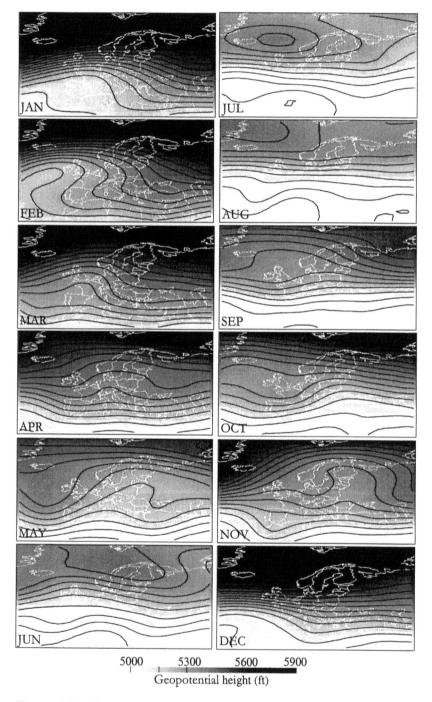

Figure 7.3 Monthly geopotential height-500 in 1993

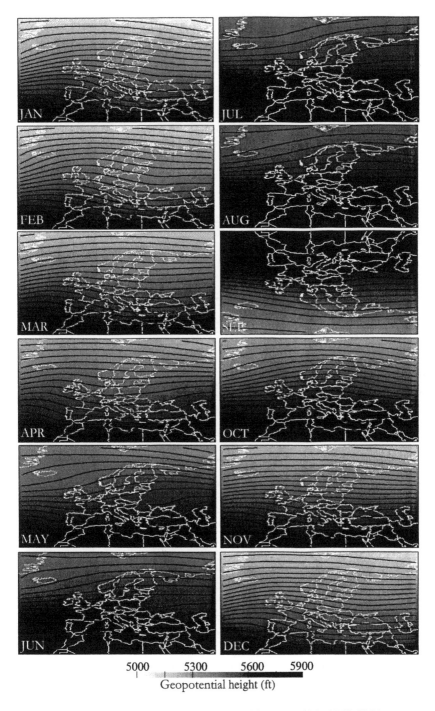

Figure 7.4   Mean monthly geopotential height-500, 1961-1990

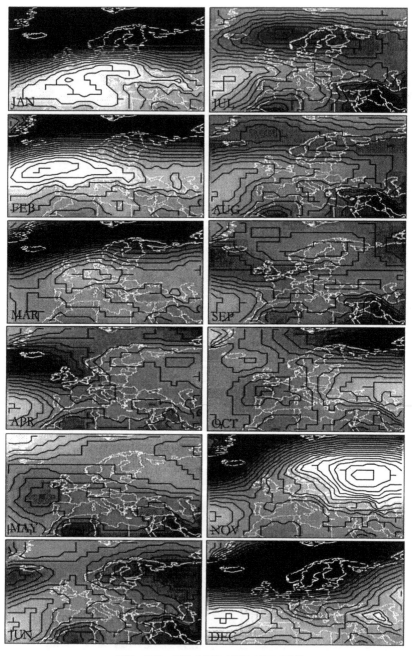

970   982   994   1005 1017 1029 1040
Sea surface pressure (mb)

Figure 7.5   Monthly sea level pressure in 1993

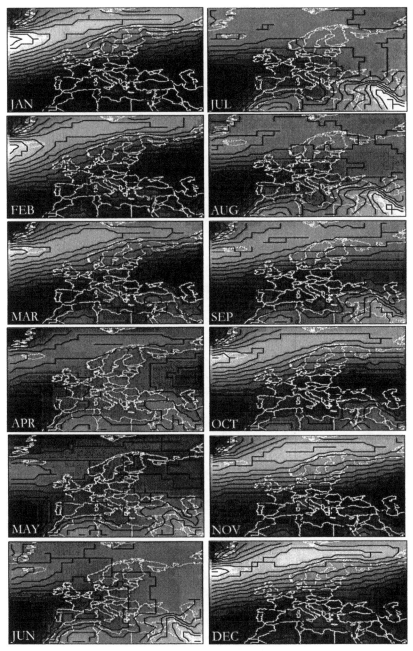

970   982   994   1005 1017 1029 1040
Sea surface pressure (mb)

Figure 7.6   Mean monthly sea level pressure, 1961-1990

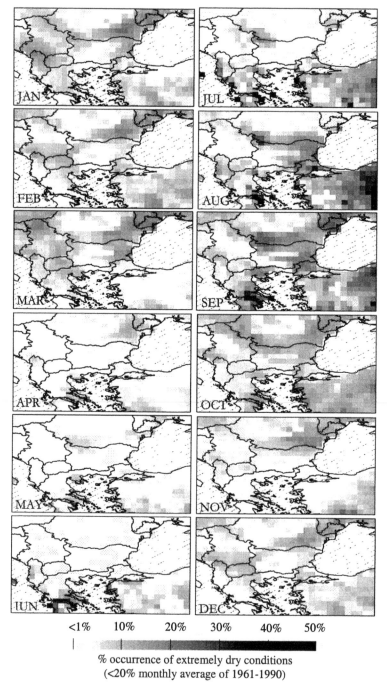

Figure 7.7 Monthly temporal and spatial variability of drought over Bulgaria, 1901-1995

PART IV
DROUGHT IMPACTS ON
WATER RESOURCES

Chapter 8

Water Resources During the Drought

Strahil Gerassimov, Marin Genev, Elena Bojilova, Tatiana Orehova

The objective of this chapter is to evaluate the quantitative and qualitative conditions of water resources in Bulgaria during the 1982-1994 drought and to use this period as a model for future climate change. This investigation is based mainly on quantitative analysis; historical data is abundant for river discharge, precipitation, and air temperature. These data correlate with global data from other countries and with astronomical parameters such as solar activity and radiation. Such correlations allow objective assessments to be made for a long historical period. Qualitative analysis is mainly used when estimating anthropogenic impacts.

Hydrological and Related Environmental Information

Data on solar activity (characterized by the number of solar spots, or "Wolf" numbers) are available from 1700 to the present, for about the last 300 years (Waldmeick 1961). The data for global temperature anomalies are available from 1854 to the present, and precipitation data for England and Wales are available from 1766 to the present (*Trends '93* 1994).

In Bulgaria, precipitation and air temperature data are available from 1892 to the present, river water levels from 1909 and systematic water discharge measurements from 1936. The number of meteorological and hydrological stations grew in the early 1950s and 1960s, but has decreased in the last two decades. Because of this, the information availability is rather non-homogeneous—there are large amounts of data for a few decades and insufficient data in the beginning and end of the hydrometeorological observations. It is necessary to consider anthropogenic factors on the non-homogeneity of the data for river discharge; after 1950 the main hydro-technical construction occurred, including water reservoirs, pumping stations, channels, hydropower stations, and irrigation and drainage systems.

The water regime in Bulgaria is described in territorial and temporal units in relation to climate, hydrographic conditions and natural chronological variations. The three main territorial units are:

1.   Danube hydrological zone: includes all Bulgarian tributaries to the Danube River (45% of the territory of Bulgaria), characterized by a temperate climate (European-Continental);
2.   Black Sea zone: includes all Bulgarian tributaries with direct discharge to the Black sea (13.8%), characterized by a sea-tempered continental climate with Mediterranean climate south of the Stara Planina mountain;
3.   Aegean hydrological zone: includes all South Bulgarian rivers with direct discharge to the Aegean Sea via the territory of Greece and/or Turkey (Struma, Mesta and Maritsa Rivers), 41.2% of the territory of the country. This zone is characterized by transition zones from temperate continental to Mediterranean (Continental-Mediterranean) climates in a north–south direction.

   Investigating river discharge variations with respect to large territorial units enabled us to avoid errors (anthropogenic effects). River runoff was assessed as the total runoff at river estuaries or country boundaries of Bulgaria for minimum temporal intervals of one calendar year. The total registered discharge of groundwater flow from ephemeral and seasonal rivers of the Dobroudja region to the Danube River and Black Sea was also investigated. The large river intakes which supply water to the Black Sea coastal cities and towns (Varna, Devnya, and Burgas) from the Kamchia and Tsonevo reservoirs were also analyzed. Some small river transfers between the Danube and Aegean Basins were not considered (but see Chapters 18 and 20 for their political significance). Evapotranspiration from irrigation systems was not taken into consideration due to lack of reliable information for irrigation schemes; in any case, irrigation systems largely collapsed in the post-1989 period. No correction was made for restoration of flows in individual river basins from reservoir operations. An example is the Arda River, a tributary that joins the Maritsa River in the territories of Greece and Turkey. It is assumed that water regulation (mostly seasonal reservoirs) and transfer from one calendar year to the next in large hydrographic zones do not affect the total annual runoff (Gerassimov *et al.* 1997; Gerassimov and Bojilova 2003).

   The total annual runoff for the three main hydrological zones for the 1960-1996 period was thoroughly investigated (based on non-homogenous data for large periods). This investigation is used as a model for all other investigations that have insufficient information. The data from hydro-meteorological stations in the estuary zones and boundary regions were used. These stations include 16 hydrometeorological stations for the Danube Basin, 10 stations for the Black sea region, and 18 stations for the Aegean zone.

   Time series from representative hydrometeorological stations with the longest observation periods, high-quality data, and minimal anthropogenic impacts were used for assessing the total river runoff for the three main

Table 8.1 Number of annual data sets with rainfall and air temperature (data from M. Genev)

Drainage Basin	For precipitation (P)	For air temperature (T),
Danube	134	71
Black Sea	52	29
Aegean	114	69
Total for Bulgaria	300	169

hydrological zones. For the period before 1936, the hydrological annual data series for some selected stations were extended using rating curves (Q = a function of gauge height, h). Comparisons with precipitation and air temperature were used. As a result of this investigation, a long data series for 16 hydrometeorological stations was obtained.

The runoff for the stations mentioned above was recorded in relative units in relation to discharge areas (runoff depth in mm). Runoff was averaged within major hydrological zones and corrected for the standard estimation (h) for the period 1960-1996. The comparison of the two types of assessments shows high reliability and confirms that data from the Black Sea basin with fewer of stations are valid.

The data series for the precipitation and air temperature for the main territorial zones were obtained from all rain gauges and meteorological stations. The number of time-series used is shown in Table 8.1. The stations for precipitation and air temperature are relatively evenly distributed over the territory of Bulgaria. The average area for precipitation covered by one station is a grid square of 19.24km. For air temperature the grid is 25.6km. The average values for precipitation (P, mm) and air temperatures (T, °C) for the three main hydrological zones for every year were obtained using linear averaged interpolation for their respective mean altitude above sea level (H, m) (Genev and Gerassimov 1998).

Analysis of Multiannual Variation in Runoff

The objective of analyzing the chronological variations in river runoff and elements of the water balance, precipitation and evaporation, is to generate possible scenarios of global climate change. In some cases additional information was used from the National Hydrological Network. Comparing trends in chronological variations of precipitation and air temperature with river runoff also makes it possible to estimate additional losses from evapotranspiration in irrigation systems, although these losses are ignored here. The results of these investigations allow us to propose some possible scenarios for the future conditions of water resources in Bulgaria. In the preliminary analysis of annual river runoff data, regression and correlation analyses between annual and total chronological values of discharge, precipitation, and air temperature for the three main hydrological zones were used. As a result, systematic errors were eliminated. Time series were extended back to 1892 (Gerassimov *et al.* 2001).

Figures 8.1 through 8.3 show annual values of solar activity (number of Wolf - **W**), solar radiation (**Si**, S-1366.9 W/m²; see Foukal 1990), temperature anomalies in the Northern Hemisphere (**dT**; *Trends '93* 1994), precipitation over England and Wales (**P-GB**), precipitation, air temperature, and runoff depth over Bulgaria (**P-BG**, **T-BG** and **h-BG,** respectively). The respective linear trend lines and fifth-degree polynomial trends show:

1. During the observation period of sunspots since 1700 (Wolf's number), an increasing linear trend of solar activity was observed;
2. The same increasing linear trend was observed for solar radiation and temperature anomalies over the Northern Hemisphere;
3. Precipitation over England and Wales also shows positive trends because of the proximity to the Atlantic Ocean. A similar correlation between solar radiation and precipitation with time delay for the Pacific Ocean circulation was obtained from the American hydrologist Perry (1992, 1994) for the western coast of North America;
4. The air temperature over Bulgaria since 1890 shows modest positive trends in accordance with temperature over the Northern Hemisphere;
5. Precipitation and river runoff for Bulgaria (since 1890) show negative trends; i.e. they are in an inverse phase with precipitation in England.

Figure 8.4 provides chronological graphs of air temperature, precipitation, and runoff depth for the three main hydrological zones. The 107-year period included 10 full sun-cycles; the tendencies are for increasing average annual air temperature for the three main hydrological zones and a decrease in precipitation and river runoff in the Danube and Aegean hydrological zones and an increase of both in the Black Sea zone.

In Table 8.2, the coefficients (a,b) of the linear trend equations $(y=ax+b)$ for the three parameters, temperature *(T, °C)*, precipitation *(P, mm)* and river runoff *(h, mm)*, are shown, where x is the number of years between 1890 and the present ($x = 0$ for 1890). The results in Table 8.2 indicate:

1. The strongest decrease in precipitation and river runoff occurred in the Danube hydrological zone (North Bulgaria);
2. In the Aegean hydrological zone (South Bulgaria), the decrease in precipitation and river runoff is weaker;

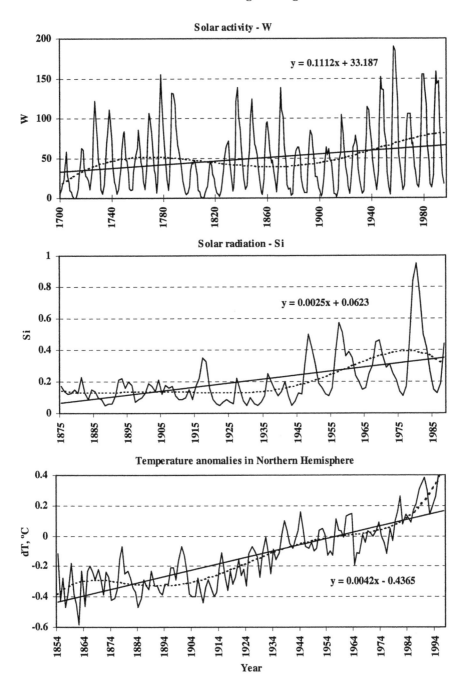

Figure 8.1 Annual values of solar activity (number of Wolf - W), solar radiation (Si), temperature anomalies in the Northern Hemisphere (dT) and respective linear and polynomial trends (power 5)

Drought in Bulgaria

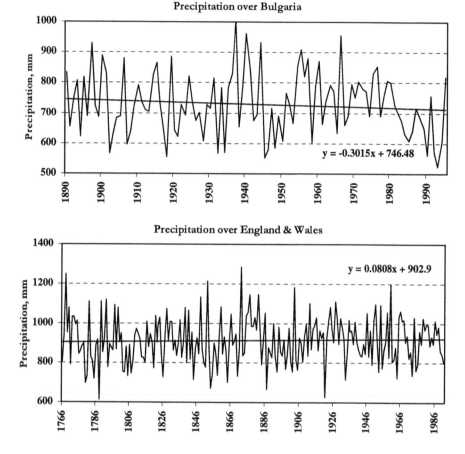

Figure 8.2 Chronological graphs of the annual values of precipitation over Bulgaria, precipitation over England & Wales and linear trends

3. In the Black Sea zone, precipitation and river runoff have modest positive trends;
4. Changes in humidity (*P, h*) in Bulgaria occur mainly from west-northwest to east-southeast; this result corresponds with the main direction of atmospheric circulation and humid Atlantic air;
5. The neutral zone with zero gradients might be located near the boundary between the European-Continental and Continental-Mediterranean climatic zones in Bulgaria. In the *Climatic Atlas of Bulgaria* this boundary line is situated along the Black Sea coast and Southeast Bulgaria from northeast to southwest.

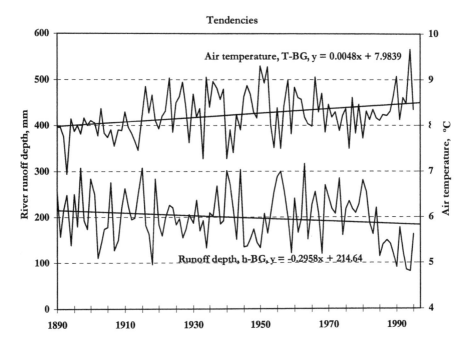

Figure 8.3 Annual values for air temperature and runoff (depth in mm) for Bulgaria, with respective linear trend lines

Defining the Drought Using Runoff and Precipitation

After 1980 Bulgarian precipitation continued to decrease while air temperatures increased. These trends led to decreases in river runoff. During the period 1982-1994 *(n=13 years)* both runoff and precipitation values were below average. During the long wet period 1954-1981, precipitation and runoff were above average. In fact, runoff and precipitation values were below average for separate instances of only one or two years during the earlier period (Bojilova *et al.* 2002).

In Table 8.3 the chronological structure of the drought period 1982-1994 is presented in relative units; the runoff and precipitation norms for Bulgaria and for the three main hydrological zones (1982-1994) are divided by the values of 106-year norms. For completeness, data for the year 1995 are presented in the last column as well. The year 1995 marks the beginning of an increasing trend in humidity and possibly the beginning of a new wet cycle (this tendency has been interrupted by several dry years).

During 1981 the precipitation over the country showed a decreasing tendency, 1% below average, but the runoff was still 29% above average. This

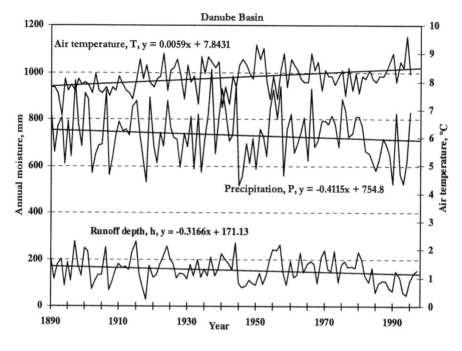

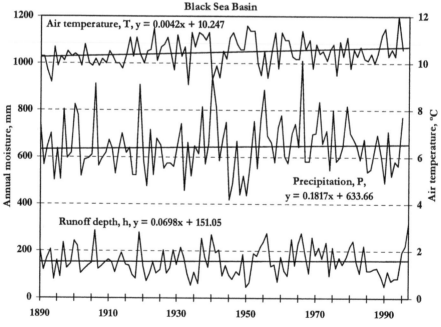

Table 8.2 Linear trends for temperature (*T*, °*C*), precipitation (*P*, *mm*) and runoff (*h*, *mm*) in the three main hydrological zones

Drainage Basin	Temperature *T*, °*C*		Precipitation *P*, *mm*		Runoff depth *h*, *mm*	
	a_T	b_T	a_P	b_P	a_h	b_h
Danube	0.0059	7.8431	-0.4115	754.8	-0.3166	171.13
Black Sea	0.0042	10.247	0.1817	633.66	0.0698	151.05
Aegean	0.004	7.4189	-0.349	773.47	-0.2929	270.51
Bulgaria	0.00481	7.9840	-0.3015	746.48	-0.2526	213.08

one-year delay in the behavior of runoff also occurred at the end of the drought period: the precipitation in 1995 was 12% above average, but the long lasting drought, especially during 1993-1994, made it possible for below-average runoff values (see Genev 1998 and 2003).

In the third column of Table 8.3 the values of the lower quartile (25% probability of non-exceedance or 75% probability of exceedance) can be validly used to delineate drought periods. Based on these parameters and river runoff values of the Danube Basin, Table 8.3 shows another 13-year drought

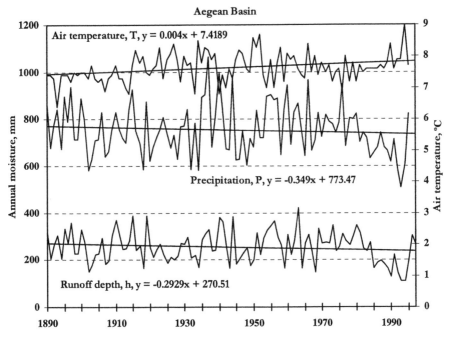

Figure 8.4 Annual air temperature, precipitation and river runoff (depth in mm) for the Danube, Black Sea (opposite) and Aegean hydrological zones and respective linear trend lines

period during 1983-1995. The lower quartile values of the runoff in the Aegean Basin and in all of Bulgaria show an 11-year drought period during 1985-1995. According to the quartile parameters, the drought period in the Black Sea zone was only six years long—from 1989 to 1994. This shortening of the drought period from north to south and especially from west to east is probably related to the availability of spatial waves with opposing phases of atmospheric circulation in the most important direction of humidity transfer to Bulgaria, from west-northwest to east-southeast.

From Table 8.3 we may conclude that the shorter, more severe periods of drought (based on runoff) most detrimental to the whole country are the periods 1983-1994 (*n=12 years*) and 1985-1994 (*n=10 years*). Within the 14-year period 1982-1995 there are two weak rises in precipitation and runoff: during 1984, when the highest runoff value for the Black Sea zone is 44% above average, and during 1991, when the highest value for the precipitation in the Danube zone is 13% above average.

In addition to investigations of river runoff and precipitation, the analysis of variations of groundwater during the drought period was considered. Examples of springs and wells from the National Hydro-Geological Network were used (Bojilova and Orehova 2000). Average values were obtained for the 37-year period (1960-1996) and for shorter periods: *n=23 years* (1960-1981), *n=13 years* (1982-1994) and *n=10 years* (1985-1994). The percent deviations of river runoff for the longer period and shorter periods were compared (Table 8.4). Figure 8.5 shows dynamics of ground-water and surface water for three main drainage zones and representative springs respectively. In general, we note that the deviations of spring discharge and groundwater levels have the same sign as the river runoff, with deviations in the same range (Orehova and Bojilova 2001). The drought

Table 8.3   Chronological structure of the drought period in relative units (*K*) for river runoff (*h*) and precipitation (*P*) with comparison to 106-year norms ($\bar{h}, \bar{P}$), K = X_K/X_{mean}

Basin	* **	Years													
		82	83	84	85	86	87	88	89	90	91	92	93	94	95
Danube *h*	.71	.85	.65	1.05	.38	.64	.71	.67	.49	.42	.96	.79	.38	.31	.69
P	.89	.91	.89	.84	.79	.86	.98	.94	.88	.71	1.13	.77	.71	.84	1.13
Black *h*	.72	1.00	.71	1.44	.76	.75	.81	.84	.60	.34	.73	.46	.55	.55	1.30
Sea *P*	.87	1.01	.92	1.05	.84	.85	.97	1.08	.93	.76	1.10	.81	.90	.88	1.20
Aegean *h*	.82	.97	.94	1.08	.65	.74	.77	.72	.65	.50	.89	.57	.43	.44	.80
P	.89	.99	.95	.84	.87	.90	.98	.90	.88	.82	.94	.78	.67	.81	1.09
Bulg. *h*	.80	.94	.82	1.11	.58	.71	.75	.72	.59	.46	.90	.63	.43	.41	.81
P	.90	.96	.92	.87	.83	.88	.98	.94	.89	.77	1.04	.78	.72	.83	1.12

* Element; ** Boundary quartile 25%.

Table 8.4 Deviations of average values for spring discharge and river runoff for the three main hydrological zones, 1960-1981, 1982-1994 and 1985-1994, in relation to their 37-year values

Place	1960-1996 $\varepsilon,\%$ **	1960-1981 $\varepsilon,\%$	1982-1994 $\varepsilon,\%$	1985-1994 $\varepsilon,\%$
Karst spring - Zlatna Panega		13.3	-21.4	-21.8
Runoff - Danube basin	-6.0	20.5	-32.3	-39.0
Karst spring - Kotel town		11.7	-20.0	-24.3
Runoff - Black Sea basin	4.9	14.8	-29.9	-38.9
Karst spring "Jazo" - Razlog		14.7	-23.2	-26.2
Runoff - Aegean basin	-3.9	14.6	-25.3	-34.0
Bulgaria – total river runoff	-3.9	17.0	-27.7	-35.8

* The analyses were made by T. Orehova and E. Bojilova.
** In relation to the period 1890-1995 (106 years).

period 1982-1994 and especially the shorter period 1985-1994 were characterized by significant lowering of groundwater. Trends in groundwater decrease are similar to those of river runoff. In the Danube and the Aegean Basins the hydrogeological drought is shown throughout 1992-1994. For the Black Sea zone major drought occurred in 1989 and 1994.

Future Runoff Scenarios

Based on the work cited above, we may calculate century-long trends ($\overline{P}, \overline{h}$) for the years 1900, 1950, 2000, 2050 and 2100. The results are presented in Table 8.5, which also shows values of the difference $P - h = E$. These values show evaporation losses to the atmosphere.

Variations in average precipitation, runoff, and evaporation for the three main hydrological zones and for Bulgaria for 100-year intervals are shown in Table 8.5. The first 100-year period is based on available data, and the second one is a forecast based on the assumption that factors such as solar activity and radiation, atmospheric circulation, and anthropogenic impacts will remain constant. The most important determining factor may be the content of greenhouse gases in the atmosphere, assuming that no actions will be undertaken to decrease them. It is possible that decreasing trends in precipitation and runoff will accelerate in the next 100 years. Estimates for the future 100 years based on trends of the past 100 years may be used as a

Table 8.5 Water balance using trend estimates for two centuries

Drainage Basin	Elements	1900	1950	2000	2050	2100	$\dfrac{\Delta y}{\Delta P}$	Δy mm
Danube	P_D, mm	751	730	710	689	668	1	-83
	h_D, mm	168	152	136	120	105	0.759	-63
	$\alpha = h_D / P_D$	0.224	0.208	0.192	0.175	0.157		
	$E_D = P_D - h_D$	583	578	574	569	563	0.241	-20
Black Sea	P_B, mm	635	645	654	663	672	1	37
	h_B, mm	152	155	159	162	166	0.378	14
	$\alpha = h_B / P_B$	0.239	0.241	0.243	0.245	0.247		
	$E_B = P_B - h_B$	483	490	495	501	506	0.622	23
Aegean	P_E, mm	770	753	735	718	700	1	-70
	h_E, mm	268	253	238	224	209	0.843	-59
	$\alpha = h_E / P_E$	0.348	0.336	0.324	0.312	0.298		
	$E_E = P_E - h_E$	502	500	497	494	491	0.157	11
Bulgaria	P_{BG}, mm	743	728	713	698	683		-60
	h_{BG}, mm	211	198	185	173	160	0.850	-51
	$\alpha = h_{BG} / P_{BG}$	0.283	0.272	0.260	0.247	0.234		
	$E_{BG} = P_{BG} - h_{BG}$	532	530	528	525	523	0.150	-9
	$W_h, 10^9 m^3$	23.42	21.98	20.53	19.20	17.76		-5.66

pessimistic scenario for possible changes in water resources and elements of the water balance. Should the trends for the next century see the norms for 2050 and 2100 similar to the values of 1950 and 1900, respectively, this situation could be used as an optimistic scenario. These results can be compared with Chapter 3. Should the magnitude of short-term deviations experienced in the 1982-1994 drought be imposed on the extrapolated trends shown in Table 8.5, water resources would be severely affected.

Global and regional models of atmospheric circulation and green house gases often use a level of $2xCO_2$. Therefore, this level of concentration is applied to the 21st century compared to the climatic period 1961-1990. From the results of the GFDL-T model often used in Bulgaria (see Sharov *et al.* 2000), the decrease in precipitation in 2060 for North and South Bulgaria was determined to be −7% and −10%, respectively. According to the same authors, the temperature increase is about +3.9°C. This value is very high, so

Figure 8.5 (opposite) Ground and surface water dynamics, 1960-1997, for three major river basins and representative springs

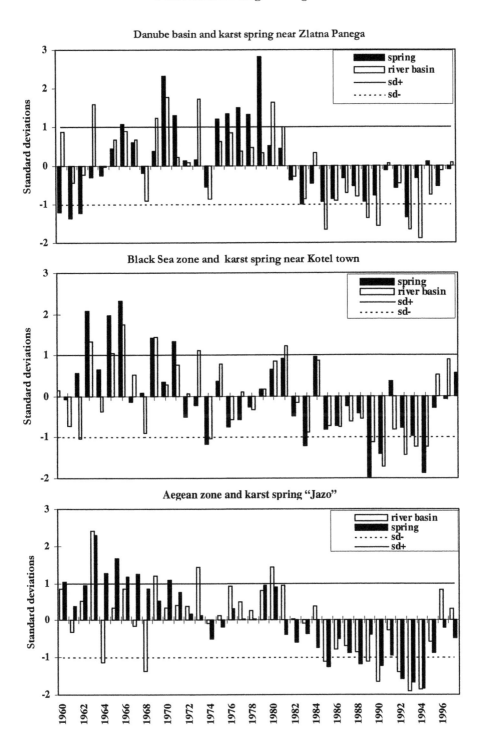

Danube basin and karst spring near Zlatna Panega

Black Sea zone and karst spring near Kotel town

Aegean zone and karst spring "Jazo"

Sharov and colleagues refer to other studies (e.g., the IPCC) which conclude that for the next century we can expect an increase in global temperature of "only" +2°C. So, we can estimate that in the next century we can expect the increase in temperature to be at least +1°C.

We determined decreases in river runoff in the Danube and Aegean hydrological zones of -28% and -34%, respectively. Hydrological analyses of climatic changes in Europe used by Arnell (1999) were simulated on the basis of four scenarios (UKHI, UKTR, CCC and GHGx). According to those models, we can expect a decrease in the river runoff for Bulgaria to be in the range of -25% to -50%, which we hope will not occur.

Conclusions

The following conclusions can be drawn from our analysis:

1. Since 1980, long-lasting decreases in precipitation, combined with increases in air temperature observed in Bulgaria, led to significant decrease in river runoff and groundwater flow;

2. During the 1982-1994, runoff and precipitation in Bulgaria were below normal. This period was characterized by a 31% decrease in runoff compared to the norms during the period 1890-1996. The relative decreases from the trend are most extreme in the Danube hydrological basin (-31%) and the lowest in the Black Sea zone (-29%);

3. The drought period 1982-1994 was preceded by the long wet period 1954-1981, when only some years were below the norm. This wet period is included in the 106-year mean (1890-1995) with overall negative trends in precipitation and runoff in Bulgaria;

4. The long 108-year period 1890-1997 is characterized by the following tendencies in the three main hydrological zones:
    a. Increases in the average annual air temperature for the three zones in accordance with the trend lines for the solar activity and radiation and temperature anomalies in the Northern Hemisphere;
    b. Decreases in precipitation and river runoff in the Danube and Aegean zones and increases for both in the Black Sea zone, in accordance with asynchrony and synchrony with precipitation over England and Wales;

5. The negative deviations in the drought period were most prominent during 1990, 1993 and 1994 when the absolute minimums of the longer period, 1890-1995, are observed. The values were from -0.31 to -0.43 of the norm.

Recommendations

To diminish the negative impacts of possible future droughts, wise use and management of available water in reservoirs, natural lakes and groundwater reservoirs, even during wet periods, is obligatory. Legislative and economic policies and incentives must promote economical water use by all consumers. As an extreme measure to avoid water crises, a policy of restrictions toward different groups of water users must be developed. When all alternatives for water conservation and other efficient management decisions have been exhausted, only then might the building of dams and interbasin transfers be considered after extensive hydrological, ecological and socio-economic impact analyses are done followed by broad public discussion.

References

Arnell, N. 1999. "The Effect of Climate Change on Hydrological Regimes in Europe: A Continental Perspective," *Global Environmental Change* 9:5-23.

Bojilova, E. and T. Orehova. 2000. "Influence of Drought Period of 1982-1994 to the Groundwater Regime in the Danube Hydrological Zone," in *Proceedings of 20th Conference of the Danube Countries*, The Slovak Republic.

Bojilova, E., S. Gerassimov, T. Orehova, M. Genev. 2002. "Natural State of Development of the River Runoff in Bulgaria Considering Anthropogenic Impact." 15 p., ICHE-2002 Conference, 18-21 September 2002. Warsaw, Poland (CD-ROM).

Climatic Atlas of Bulgaria. 1956. Sofia: Technika.

Foukal, P. and J. Lean. 1990. "An Empirical Model of Total Solar Irradiance Variation Between 1874-1988," *Science*, 24(4):556-559.

Genev, M. and S. Gerassimov. 1998. "Tendencies in the Multi-Annual Variation of River Discharge in Bulgaria," *Scientific Report*, Library of the Bulgarian Ministry of Education and Science, N: 406, Volume III, pp. 77.

Genev, M. 1998. "On the Probabilistic Character of the Phenomenon River Discharge. *Journal of the Bulgarian Academy of Sciences* 3-4:43-48.

Genev, M. 2003. "Patterns of Runoff Change in Bulgaria," in *Water Resources Systems-Water Availability and Global Change* (Japan, Sapporo, 10-11 July). Sapporo, *IAHS, Publication* 280: 175-185.

Gerassimov, S. and E. Bojilova. 2003. "Assessment of Capacity and Tendencies of Bulgarian Water Resources," *Journal of the Bulgarian Academy of Sciences* 1/2003:9-18.

Gerassimov, S., E. Bojilova, T. Orehova, M. Genev. 2001. "Water Resources in Bulgaria During the Drought Period – Quantitative Investigations," in *Proceedings of the 29th Int. Association of Hydraulic Research and Engineering Congress*, September 2001, Beijing, China.

Gerassimov, S. *et al.* 1997. "Water Resources and Hazards," pp. 199-228 in C. G. Knight *et al.* (editors), *Global Change and Bulgaria*. Sofia: National Coordination Center for Global Change.

Orehova, T. and E. Bojilova. 2001. "Impact of the Recent Drought Period on the Groundwater in Bulgaria," in *Proceedings of the 29th International Association of Hydraulic Research and Engineering Congress*, September 2001, Beijing, China.

Perry, C. A. 1992. "A Correlation Between Precipitation in the Western United States and Solar Irradiance Variations," pp. 721-729 in *Proceedings of the American Water Resources Associations*

Conference Managing Water Resources During Climate Change, Reno, Nevada, 1-6 November 1992.

Perry, C. A. 1994. "Solar-irradiance Variations and Regional Precipitation Fluctuations in the Western USA," *International Journal of Climatology* 14:969-983.

Sharov, V. *et al.* "Climate Variability and Change," pp. 55-96 in C. G. Knight *et al.* (editors), *Global Change and Bulgaria.* Sofia: National Coordination Center for Global Change.

Trends '93. 1994. *Trends '93, A Compendium of Data on Global Change.* Carbon Dioxide Information Analysis Center, Oak Ridge National Laboratory, USA.

Waldmeick, M. 1961. *The Sunspot Activity in the Years 1610-1960.* Zurich: Schulthess & Co. AG.

Chapter 9

The Impact of Drought on Surface Water Quality

Galia Bardarska, Hristo Dobrev

Background

The term "water quality" describes the physical, chemical, and microbiological characteristics of water. Surface water quality depends on a complex system of inputs and feedbacks. Superimposed on natural chemistry are factors like catchment land use, deposition of atmospheric pollutants, discharge of urban and industrial wastewater, drainage from urban areas, and accidental pollution. The effects of drought on surface water quality depend very much on local environmental conditions as well as on legislative and economic pressures, such as water quality targets for individual rivers and the price of licenses for effluent discharges.

In this chapter, the reservoir water in protected areas is investigated. These areas were selected so that the impact of human activity on water quality in the drought period 1982-1994 could be minimized. The main message to decision-makers is that with the use of appropriate technological schemes and chemical products, polluted reservoirs could provide a safe and sustainable source of water during drought periods.

Existing Regulatory Framework for Assessment of Water Quality

In the period 1982-1994, most water quality assessments and regulations were primarily based on two documents, which include Regulation No. 7 on flowing surface waters issued in 1986 by the Committee on Environmental Protection in the Ministry of Public Health and the Committee on Regional Construction and Spatial Planning, and Bulgarian State Standard BSS 2823-83 for "Drinking Water." According to Regulation No. 7, the categorization of rivers in the Republic of Bulgaria is based on the quality of river water measured in 1967 and amendments to the Regulation introduced in 1985. Table 9.1 shows the main indices used in determining the three categories of

Table 9.1 Main indices for determining surface water quality according to Regulation No. 7/1986

Indices	Flowing surface water categories		
	Category I	Category II	Category III
1 Iron, mg/l	0.5	1.5	5
2 Manganese, mg/l	0.1	0.3	0.8
3 Dissolved oxygen, mg O_2/l	6	4	2
4 Oxidizability (KMnO$_4$), mg O_2/l	10	30	40
5 BOD$_5$, mg O_2/l	5	15	25
6 Ammonium, mg/l	0.1	2	5
7 Phosphorus, mg/l	0.2	1	2
8 Suspended solids, mg/l	30	50	100

rivers as they are currently defined. Category I refers to waters suitable for drinking water. Those belonging to Categories II and III are generally acceptable from an ecological point of view, but are not of sufficient quality to be used for potable water. The categories of surface waters with respect to the values of their quality indicators differ from European standards. For instance, according to European directives, the values of the indicator Biological Oxygen Demand in 5 days (BOD$_5$) are as follows: Category I – 3 mg O_2/l; Category II – 5 mg O_2/l; and Category III – 7 mg O_2/l.

Within Bulgaria, qualitative monitoring of surface waters is primarily carried out by the Regional Inspectorates of Environment and Water, the Executive Environmental Agency within the Ministry of the Environment and Water, the National Institute of Meteorology and Hydrology, and the Forest Research Institute with the Bulgarian Academy of Sciences. Responsibility for the control of potable water lies with the Inspectorates of Hygiene and Epidemiology, the Water Supply and Sewerage Companies, and the National Center of Hygiene, Medical Ecology and Nutrition.

BSS 2823-83 "Drinking water" addressed 39 specific water quality indicators. The most frequent deviations of raw waters occur with measurements for turbidity, oxidizability, iron, manganese, ammonium, nitrites and nitrates, as well as for phytoplankton and microbiological indicators. The Bulgarian standard for drinking waters did not address all of the pollutants monitored by the World Health Organization (1993) nor those considered in European standards (1998).

With respect to potable water supplies, the control of compounds formed when surface water is treated with different reagents and disinfectants is of particular importance. For instance, when aluminum-containing reagents are applied, the content of residual aluminum should not exceed the recommended and obligatory norms of 0.05 mg/l and 0.2 mg/l, respectively (Holdsworth 1991). When potable water is disinfected with chlorine products, the quantity of trihalomethanes should also be controlled (Holdsworth 1991; Dore 1989). Obligatory control of these two parameters, in addition to the indicators that are subject to permanent control according to BSS 2823-83, is indispensable, especially when water pollution increases during periods of drought and larger quantities of chemical products are needed to assure safe water supplies.

Existing Infrastructure and Related Problems

About 98.7% of the Bulgarian population is connected to a centralized water supply network. During the period under review (1982-1994), the relative percentage of human settlements served by centralized potable water systems increased from 80% to 84.7% of the total number of human settlements in the country. Of 9,688 potable water sources, 39% are of the gravity type and 61% are of a pumping-type. Forty-six potable water treatment plants (PWTPs) have single-layer rapid filters for physical-chemical treatment with aluminum sulfate and gaseous chlorine for disinfection. Some also have horizontal sedimentation tanks. Additionally, ozonation processes for disinfection are in operation at potable plants in Breznik and Kardjali. One exception to this typical approach is the process used in the two PWTPs in towns of Targovishte and Preslav. In these plants, only micro-sieves, ozonators, and chlorinators are used for treatment of water that is supplied by the Ticha Reservoir, which was designed for irrigation purposes. The applied treatment system has no specific facilities for retention of mechanical substances and oxidized organic compounds.

Under normal conditions (those periods without torrential rainfall or prolonged drought), the majority of potable water treatment plants deliver drinking water in compliance with BSS 2823-83. In the case of small human settlements without potable water treatment plants, direct application of $Ca(OCl)_2$ or $NaOCl$ is typically utilized for disinfection of natural waters.

Under certain conditions, the water supply network itself can act as a source of additional pollution of the water supplied to consumers. Approximately 77.36% of the 22,388km of main water delivery pipe systems and 81.38% of the municipal water supply network within Bulgaria are made of asbestos-cement. Breakage of these asbestos-cement water pipelines (most of which were laid down between 25 and 30 years ago) can increase at a particularly high rate as a result of the frequent switching on and off of water supplies during periods of water rationing (regimes). Such breakdowns usually result in long-term pollution of the water distribution networks and changes in the quality of the supplied water. In an assessment of the state-of-repair of the Water Supply and Sewerage Company in the city of Sofia in 1992, the French company SAUR found that disruptions of the water distribution network not only had a negative effect on water quality but also added considerable losses to the water reserves in the Iskar Reservoir. This exemplifies the aggravating conditions that occur in periods of drought and water rationing (SAUR 1992).

In the majority of cases when deviations between monitored water quality and defined water criteria are observed, either urgent amelioration measures are taken or the water supply is purposely interrupted. Due to the fact that most urban areas have centralized water supply systems, there is much more concern for rural areas where local water supplies are obtained from wells and other local water sources without sanitary protection. In these areas, the existence of septic pits and the application of natural fertilizers are cited as reasons for both the increased content of ammonium compounds and phosphates, as well as the increase in the number of microbes in groundwater. Similar problems are also encountered in the case of wells for centralized water supply systems that do not have sanitary protection zones. For instance, the Bivolare watershed near the town of Pleven has been polluted by an urban waste water collector (DEPA 1998).

Incomplete or inadequate sewerage systems throughout the country are another source of permanent anthropogenic pollution of water sources. As shown in Table 9.2, only 35.7% of the population is connected to wastewater treatment technologies (National Statistical Institute 1998).

The poor quality of aeration systems in the case of biological wastewater treatment, and the absence of facilities for treatment of separated sludge, are also reasons for disruption of the operational duty cycle of the plants and for the discharges of untreated wastewater to water bodies in violation of Regulation No. 7. The historical use of mechanical and biological methods of

Table 9.2 Wastewater treatment plants (WWTPs)

Treatment technologies	Number of WWTPs	Design capacity, m^3/d	Working capacity, m^3/d	Connected population, %
Mechanical (first step)	13	41,452	29,747	0.9
Biological (second step)	38	1,811,326	1,153,829	34.8
Total	51	1,852,778	1,183,576	35.7

waste water treatment has not sufficiently reduced problems of heavy metals, nitrogen, and phosphorus in Bulgaria (Halcrow 1999; Ecoglasnost 1999). The control of wastewater in urban areas is further hampered by the discharge of untreated wastewater from industrial enterprises. Consequently, during extensive dry periods, increased control and application of specific measures are necessary for wastewater treatment plants and septic pits of small human settlements located in the area of sanitary protection zones of potable water sources.

The poor state of the water supply and sewerage systems of the country, as described above, indicates that in the case of diminished water resources in periods of drought, the preconditions for deterioration of water quality increase.

A Case Study on the Impact of Drought on Water Quality

The period under review (1982-1994) coincides with a period of tremendous change in the economic system of Bulgaria, in which a transition from a centrally-planned socialist economy to a market-oriented economy has taken place. Since November 10, 1989, state ownership in the areas of agriculture, forests, and industry was discontinued, resulting in a substantial reduction in agricultural activities and the closure of a number of enterprises. During this transition, the environmental situation in Bulgaria has actually shown some improvement. Figure 9.1 illustrates trends in water quality improvement in two major rivers, Maritsa and Iskar, during the period 1989-1996.

Figure 9.2 shows a similar improvement trend for the Vit River, which is located in the most polluted section of the area near the village of Bivolare (DEPA 1998). This particular stretch of river is designated as Category III, and is downstream from some of the largest sources of pollution in the area, such as a petroleum refinery and the towns of Pleven and Dolna Mitropolia, both of which have well-developed industrial and agricultural sectors.

It is evident that drought has negatively impacted the quality of flowing surface waters during the economic transition period that has occurred since 1989. The relationship between water quality and drought conditions has been exhibited by three different reservoirs: Studena, Kamchia, and Iskar (see Figure 17.1 in this book for their location). In each case, a general deterioration of water quality has been observed in association with a precipitous drop in the water level behind the dams, whereby water contamination may be facilitated by increased interaction between the water and sediments at the bottom of the reservoir. The measured concentrations

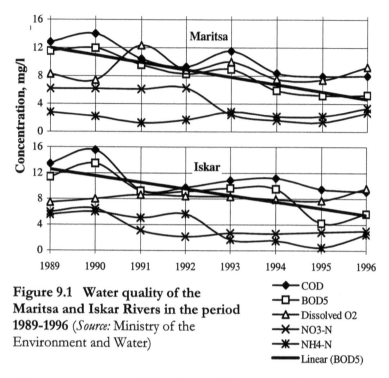

Figure 9.1 Water quality of the Maritsa and Iskar Rivers in the period 1989-1996 (*Source:* Ministry of the Environment and Water)

of iron, manganese, and aluminum of the water in the pores of the sediment of Iskar Reservoir in May 1994 were recorded as 1.2, 0.25, and 0.7 mg/l, respectively (Hrischev *et al.* 1994). The increased bottom-layer pollution also leads to reduction of the dissolved oxygen content of the water.

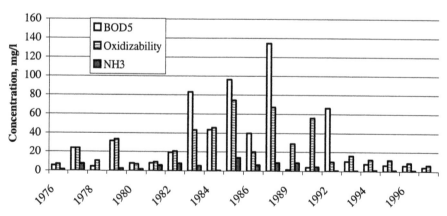

Figure 9.2 Water quality of the Vit River at the village Bivolare in the period 1976-1997 (*Source:* DEPA 1998)

Similarly, measurements of dissolved oxygen in the water of Studena Reservoir by horizons for the period of October 1991 through September 1993 showed values far below the norm of 2.0 mg O_2/l for surface water of Category III. The value of dissolved oxygen was 0.99 mg O_2/l in February 1992, 0.91 mg O_2/l in February 1993, and 0.2 mg O_2/l in August 1993 (ECO AQUA TECH 1994). During the same time period, the lowest value of dissolved oxygen in the Kladnishka and Matenitsa Rivers, which discharge to the Studena Reservoir, was measured in August 1993 at a level of 6.19 mg O_2/l in the Kladnishka River. Therefore, in terms of dissolved oxygen, all rivers meet the standards for Category I surface water (ECO AQUA TECH 1994). The highest concentrations for forms of iron and dissolved manganese for the same period have been measured at 27m below the surface of the Studena Reservoir (2.4 mg Fe/l in June 1993 and 2.32 mg Mn/l in February 1992; ECO AQUA TECH 1994).

Another indicator for deterioration of water quality is the presence of phytoplankton whose development is accelerated by temperature increases, low water levels, and the presence of nutrients. The quantity of phytoplankton cells of larger dimensions and fiber-type forms is a threat because phytoplankton may clog filters used in water treatment plants, or may pass into the water supply network.

The problems of several PWTPs in terms of their technology and operational practices during the period 1982-1994 are discussed below in parallel with the quantity and quality of water in the three reservoirs. Observations and experimental studies for the Studena and Pancharevo PWTPs provide a basis for making concrete recommendations on the use of adequate technological schemes, treatment equipment, and reagents in periods of drought.

Specific Indicators of Drought Impact on Reservoir Water Quality

Studena Reservoir: During the period 1982-1994, the volume of water in the Studena Reservoir was below 50% of its total capacity of 25.2 million m^3, and was below its "dead volume" of 2.4 million m^3 during the months December 1993-March 1994.

Removal of turbidity typically does not present a problem when a treatment plant is operating according to standard procedures, but under drought conditions, attaining of requirements for potable water standards in terms of manganese content and phytoplankton is much more difficult when using the classical method of applying aluminum sulfate as a coagulant. In the case of both low and high turbidity values, the older two-step Studena

Drought in Bulgaria

Table 9.3 Turbidity of the Studena Reservoir and treated water at Studena Drinking Station

Date	Turbidity of the Studena Reservoir mg/l	Turbidity of treated water at the new PWTP, mg/l	Turbidity of treated water at the old PWTP, mg/l
24.03.1993	30	15	2.8
25.03.1993	29	5.2	1.4
26.03.1993	32	20	2.2
28.03.1993	31	12	2.2
29.03.1993	24	7.2	1
31.03.1993	28	12.4	2.4
01.04.1993	25	12	5.2
20.10.1994	7	3.2	0.3
21.10.1994	6	2.5	0.45
22.10.1994	6.5	3	1.9
23.10.1994	7	3	2.2
24.10.1994	6.5	3.2	0.4

Source: Water Supply and Sewerage Company, Pernik municipality

treatment plant (whose equipment included sediment tanks and rapid sand filters) typically achieved better treatment results than the newly constructed one-step Studena plant featuring only rapid sand filters (Table 9.3).

Due to extremely low water levels, manganese content in reservoir water exceeded the allowed value of 0.1 mg/l for drinking water during the years 1983, 1986, and 1990-1994. Water treated with aluminum sulfate had manganese content that ranged from 0.13 mg/l (23 March 1993) up to 0.33 mg/l (29 January 1990). The worst deterioration of the hydro-biological indicators was recorded in August 1993 when the reservoir was characterized by mass development of blue-green algae. The quantity of general chlorophyll "a" (Parsons-Strickland) in the phytoplankton was also the highest in August 1993 (ECO AQUA TECH 1994).

Kamchia Reservoir: The Kamchia Reservoir (228.8 million m³ total capacity and 74.6 million m³ "dead volume") is the main water source for the towns of Burgas and Varna. During the period 1989-1990, the population was subject to a period of drastic water rationing. In December 1990 the volume of water reached 50.4 million m³ and the water quality gravely deteriorated, as exhibited by readings of 60.3 mg/l for turbidity, 4.4 mg O_2/l for oxidizability, 1.14 mg/l for iron, and a manganese content 0.65 mg/l. With increased levels of turbidity, concomitant deterioration was noted for other monitored

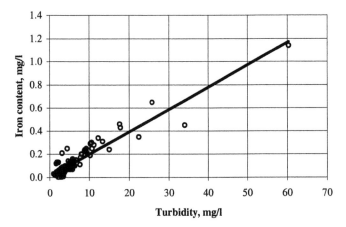

Figure 9.3   Relationship between iron content and turbidity of the Kamchia Reservoir water, August 1985-December 1998

indicators as well. For example, Figure 9.3 provides an example of the linear dependence between turbidity and iron content.

During the water rationing period, the Kamchia PWTP faced difficult operating conditions not only with respect to the provision of the required quantity of water for the population, but also in terms of the desired quality of the drinking water.

Iskar Reservoir: In December 1994, the Iskar Reservoir which has a total volume of 670 million m³ and an allowable "dead volume" of 90 million m³, was emptied to 66 million m³ water. Despite the low water temperatures of 1-2°C during the winter of 1994/1995, both the iron and manganese content and the quantity of phytoplankton increased above the allowed values of 0.2 mg/l, 0.1 mg/l and 100 cell/ml, respectively. The quantity of phytoplankton in the reservoir reached 6,349 cell/ml (14.02.1994) and 2,860 cell/ml in the treated water (14.03.1994). The relationship between the water volume of Iskar Reservoir and manganese content is shown in Figure 9.4.

In addition to the water shortage during the period November 1994-May 1995, the Water Supply and Sewerage Company of the capital city of Sofia had difficulties in the treatment of reservoir water at the Pancharevo PWTP. The imposed monthly limits on drinking water consumption by industrial enterprises (2.46 million m³ in 1994 and 1.04 million m³ in 1995) and the obligatory use of water from proprietary sources for technological applications reduced the water flow from the Iskar Reservoir below the operative capacity of 4.5 m³/s for the Pancharevo PWTP (Water Supply and Sewerage–Sofia 1994). Despite this drop in water supply from Iskar

Drought in Bulgaria

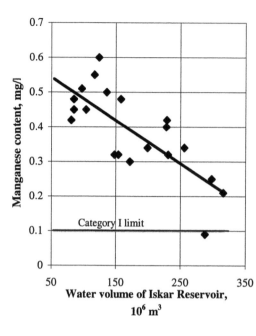

Figure 9.4 Relationship between water volume of Iskar Reservoir and manganese content in the period March 1993-April 1995

Reservoir, part of the water reached the consumers directly from the reservoir subjected only to chlorination, and water passing through the treatment plant was not treated by reagents for physical-chemical purification. For the period January 1994-April 1995 alone, the total quantity of untreated reservoir water supplied to the population amounted to 82 million m³ (Water Supply and Sewerage–Sofia 1994). The higher concentrations of chloroform in the Sofia urban water supply network compared to the values measured at the exit of the potable water treatment plant Pancharevo provides evidence that untreated drinking water had been supplied to consumers (Table 9.4; Hrischev *et al.* 1994). As a result of poor water management, pollution of the water delivery network set in, and the recorded values of a number of quality indicators of consumed water were higher than those of reservoir water (e.g., 1.4 mg/l for iron and 0.97 mg/l for manganese; Gopina and Vassilev 1998).

Despite these conditions, the supply of drinking water of poor quality to the consumers and the pollution of the water pipeline network could have been avoided if the coagulant-flocculant-sorbent CFS-SOLVO® had been used in the Pancharevo PWTP. As shown in Figure 9.5, this reagent appeared to perform well when tested on 1/6 of the equipment in the plant in November-December 1995 (Dobrev and Bardarska 1996). During the course of the investigation, a maximum water flow through the sedimentation tank (Pulsator type) under observation was maintained (2600-2700 m³/h), as was the added reagent (from 26 to 54 l/h). The dose was in the range of 10 to 20

Table 9.4   Trihalomethanes (THM) in Sofia water, May 1994

Water sample from:	Chloroform	Dichlorbrommethane	Dibromchlormethane	Bromoform	Total THM
Iskar Dam	< 1	< 0.1	< 0.1	<1	0.0
Exit, potable water station "Pancharevo"	14.7	2.7	0.4	<1	17.8
Water supply of Sofia region "Mladost"	25.3	6	0.8	<1	32.1

ml solution to 1m³ treated water. The cost of the reagent is about US$ 0.0015-0.003 per m³ treated water. As noted in this instance, even with low turbidity and temperature of reservoir water, a stable duty cycle of sedimentation in the Pulsator typically sets in quite rapidly (in this case, 9-10 hours after the initial feeding of CFS). The large and heavy floccules that are formed settle at the bottom of the Pulsator and make a stable cloud that cannot be disrupted by the upward water flow. In actual fact, the exceptionally good formation of floccules ensures the absence of residual aluminum in the treated water (from 0.00-0.04 mg/l), and has a considerable purification effect on all surveyed indicators including manganese immediately after passing through the Pulsator (Figure 9.5). Manganese elimination depends on the turbidity of raw water. Over 70% manganese elimination after sedimentation can be achieved by means of CFS application when the turbidity of the reservoir water is greater than 5 mg/l.

As a result of complaints by inhabitants of the Lyulin housing complex in Sofia about the poor drinking water quality during the water crisis, lawsuits were deposited with the City Prosecutor's Office (Ref. No. 3607/1994) and the National Investigation Office (Ref. No. 96/1995). Unfortunately, though, the cases were discontinued for lack of any laws that assign personal responsibility to those guilty of supplying drinking water of poor quality.

Conclusions

The drop in industrial and agricultural production in 1982-1994 on the whole produced some improvement in the quality of surface and ground water during the periods of drought. Still, deviations from legislated requirements for drinking water quality per Bulgarian State Standard 2823-83 "Drinking

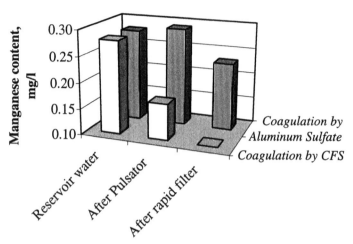

Figure 9.5 Manganese elimination by different chemical products at Pancharevo PWTP on December 3, 1994.

Water" have been observed as a result of a drop in reservoir water levels and the concomitant introduction of water rationing. Several conclusions may be drawn based on observations made for three reservoirs (Studena, Kamchia, and Iskar) during the drought period. At very low water levels, deviations from BSS 2823-83 can occur in reservoir water with respect to color, turbidity, oxidizability, iron, manganese, and phytoplankton content. The application of aluminum sulfate in plants that are in a state of technical disrepair is not adequate to ensure quality treated drinking water under adverse conditions. Thus, the supply of improperly treated surface water to the water supply network under drought conditions can lead to additional long-term pollution of the system. Currently, there is no regulatory framework that allows consumers to seek personal responsibility of decision makers who have allowed the supply of poor-quality drinking water.

Recommendations

The deterioration of drinking water quality from surface waters during periods of drought may be avoided by using a variety of strategies. Sanitary protection zones around water sources should be defined and strictly observed. Construction of two-step technological treatment schemes using effective sediment tanks and rapid filters for surface waters and application of adequate filtering materials and reagents for physical and chemical treatment should be undertaken. Water supply systems should be adequately maintained, and water rationing by interrupting supply (regimes) should be avoided—regimes do not always lead to water savings because hydraulic shocks can disrupt water distribution pipes and because vacuum is created when the water flow is stopped. Secondary compounds (trihalomethanes, chlorpicrine, residual aluminum, etc.) above allowed values are health risks that must be controlled. Consumers could implement home treatment mechanisms (boiling, tablets, filtering, etc.) for tap water during periods of drought. Finally, national legislation should hold officials responsible when drinking water of inadequate quality is provided by water supply systems in violation of national standards.

References

Committee on Environmental Protection, Ministry of Public Health and the Committee on Regional Construction and Spatial Planning. 1986. "Regulation No. 7 for Indices and Norms for Determination of Surface Water Flow," *State Gazette* 96:6-8.

DEPA. 1998. Project DEPA No.124/008-004 "Improvement of Ground Water Supply Situation and Identification of Ground Water Protection Zones for Well Fields in Pleven." Main Report of Carl Bro a/s Samfudsteknik, Denmark, pp. 69.

Dobrev, H. and G. Bardarska. 1996. "Sofia and Istanbul Drinking Water Treatments by Coagulant-Flocculent-Sorbent CFS-SOLVO," *Water Affairs*, 3/4:34-37.

Dore, M. 1989. *Chimie des Oxidants & Traitement des Eaux*. Paris: Lavoisier Tech. & Doc.

Ecoglasnost. 1999. *Environment, Health, Society*. Sofia.

ECO AQUA TECH. 1994. Report on the Topic 050/003.91.46 "Physical-Chemical and Hydrobiological Investigations of the Water of Studena Dam and Its Tributaries Struma River, Kladnichka and Matnitsa for Determination of Zone "B" of Sanitary Protection by Self Treatment Ability." Sofia.

Gopina, G. and K. Vassilev. 1998. "Estimation of the Drinking Water Quality of Sofia City in the Water Rationing Conditions—Period 1994-1995," *Water Affairs*, 1/2:22-26.

Halcrow. 1999. *Provision of Technical Assistance in the Approximation of Water Legislation for Bulgaria – BUL 110*, PHARE.

Holdsworth, R. 1991. *New Health Considerations in Water Treatment*. Aldershot, UK: Avebury Technical.

Hrischev, H. *et al.* 1994. *Theoretical Background of Eco-Geology Dam Modeling. Sedimentation-Geochemical Model of Iskar Dam, Part II*, Geological Institute at the Bulgarian Academy of Sciences.

National Statistical Institute. 1998. *Environment '97*. Sofia.

SAUR. 1992. *Assessment Factors of the State-of-Repair of the Water Supply and Sewerage Company of the City of Sofia*. Municipality of Sofia.

Water Supply and Sewerage Company-Sofia. 1994. *General Director's Report to the Sofia Mayor*. No. RD-01-6900/22.12.1994.

Water Supply and Sewerage Company-Sofia. 1995. *General Director's Report to the Sofia City Prosecutor*. No. RD-01-3077/23.05.1995.

PART V
DROUGHT IMPACTS ON
NATURAL ECOSYSTEMS

Chapter 10

The Impact of Drought on Natural Forest Ecosystems

Ivan Raev, Boyan Rosnev

Natural ecosystems have been preserved in excellent condition in the forests of Bulgaria. About two thirds of these forests are natural, primary forest ecosystems which in the past covered a much greater area in Southeast Europe; most have not been affected by afforestation activities. The impact of drought on natural Bulgarian forest ecosystems is an indicator of the potential effect of further warming and drought, resulting from global climate change during the next century. Our analysis points to measures that can be taken now which have the potential to alleviate the negative impact of global warming in Bulgaria.

Methodology and Sites

Statistical data on forests in Bulgaria are drawn from the National Statistical Institute. The data on the distribution of forests in Bulgaria according to altitudinal belts as well as precipitation over the territory of Bulgaria are taken from Kostov *et al.* (1976). The data from ecological stations of the Forest Research Institute have been used to characterize natural forest ecosystems in Bulgaria, as follows:

1. Oak forest belt - the Souvorovo Station (160m elevation) in northeastern Bulgaria (*Quercus cerris* or Turkey oak forest), Souvorovo Forestry Commission, Varna;
2. Oak and beech forest belt - the Gabra Station (850m) in the Ichtimanska Sredna Gora mountain range (*Quercus petraea [sessiflora]* and *Fagus sylvatica,* durmast oak and European beech), the Elin Pelin Forestry Commission, Sofia District;
3. Beech forest belt - the Balkanets Station (1200m) in the Sredna Stara Planina (120 year old *Fagus sylvatica* or European beech forests), Troyan Forestry Commission, Lovech Region;

Table 10.1 Statistical data on forests in Bulgaria, 1955-1995

Characteristics	1955	1965	1975	1985	1995
Total area, 10^6ha	3.67	3.51	3.69	3.77	3.77
Afforested area, 10^6ha	3.15	3.05	3.13	3.24	3.26
Percentage of conifers, %	14.00	23.50	29.70	36.10	32.70
Protected forests, %	8.40	12.20	19.00	29.20	39.80
Mean increment, 10^6m^3	6.10	5.90	6.83	9.11	12.35
Total volume, 10^6m^3	244.70	248.10	268.50	336.70	456.70
Cut (planned), 10^6m^3	6.82	6.84	6.86	6.45	6.24
Cut (actual), 10^6m^3	7.45	8.16	6.32	5.53	4.76
Produced seedlings, 10^6	414.30	598.00	637.20	351.00	156.00
Afforestation, 1000ha	34.34	40.28	48.97	29.62	9.16

4. Pine forest belt – the V. Serafimov Station (1550m) on the southern Rila Mountain slopes (90-year-old *Pinus sylvestris* or Scots pine forests), Yakorouda Forestry Commission, Blagoevgrad Region;

5. Spruce forest belt – the Ovnarsko station (1550m) on the northern Rila Mountain slopes (90-year-old *Picea abies* or northern spruce forests) Samokov Forestry Commission, Rila Mountain (Raev 1994).

   Climate and water circulation in the forests were based on data from ecological stations, including air temperature and precipitation measured daily at 07.00, 14.00 and 21.00 hours. The de Martonne (1925) aridity index was calculated [P / (T+10); P, mm; T, °C], and climatograms were drawn using the method of Walter (1973). Soil moisture was measured every fifth and twentieth day of the month over a long-term period (Raev 1989). Dendrochronological data was taken from Grozev and Delkov (1995).

The Forests of Bulgaria

From 1955-1995, Bulgarian forests have been relatively constant in area, with a trend of slow increase (Table 10.1). In 1995 their area was 3,770,000ha, which constitutes 34% of the territory of the country. The increase of coniferous species in Bulgarian forests is a characteristic feature: from 14% in 1955 they reached 37% in 1990, with a subsequent decrease to approximately 32.7% of the forest area in 1995. These changes are linked with the dynamic rate of afforestation over the 1955-1985 period, chiefly with coniferous species. A later change of national forest strategy led to an increase of

Table 10.2   Distribution of forests in Bulgaria according to tree species, areas and volume of timber, 1995

Tree species	Area	Volumes	Increments
	x 1000ha	x 1000m³	
Pinus sylvestris L.	562.73	84,508.7	3,246.5
Pinus nigra Arn.	313.64	21,934.9	997.2
Picea abies (L.) Karst.	153.87	34,583.2	545.0
Abies alba Mill.	30.53	9,984.0	125.7
Other conifers	38.30	5,649.1	162.1
Total conifers	**1,099.08**	**156,659.9**	**5,076.5**
Quercus sp.	1,286.04	99,029.1	2,676.8
Fagus sylvatica L.	477.66	101,977.3	1,610.7
Robinia pseudoacacia L.	90.28	4,918.7	411.2
Carpinus betulus L.	86.82	15,015.7	416.9
Tilia sp.	41.77	5,676.4	208.2
Populus sp.	22.32	1,771.4	186.1
Other broad-lived	155.11	12,350.7	519.6
Total broad-leaved	**2,159.99**	**240,379.3**	**6,029.5**

afforestation with broad-leaved tree species, which have greater drought resistance, resulting in the reduction of the coniferous share.

There was a doubling of the annual growth over the 40-year period from 6,100,000m³ to 12,350,000m³ in 1995, together with a parallel increase in timber stock: from 244,700,000m³ to 456,700,000m³ in 1995, also attributable to afforestation. At the same time, a fall in forest felling from 8,570,000m³ in 1960 to 4,760,000m³ in 1995 also led to an increase in the reserves of the stands. An exceptional expansion of afforestation occurred over the 1955-1980 period in Bulgaria, with up to 86,660ha afforestated area for some years. In this way about 1,200,000ha of new forests were created over a 40 year period, accounting for about 33% of the total forest area. An interesting index is the distribution of forests according to tree species. Table 10.2 (after Raev *et al.* 1996) gives data on the main tree species, together with the timber stocks.

Specific Ecological Conditions in Bulgarian Forests

Ecological conditions for forest vegetation in Bulgaria changes according to altitude as well as geographical location of the terrain, from north to south and from west to east. Table 10.3 shows the distribution of forests in Bulgaria depending on altitude, based on material by Kostov and colleagues (1976). It should be noted that 60.6% of the forests in Bulgaria are in altitudes up to

Table 10.3   Forest lands according to altitudinal belts (%)

Altitudinal belts	Altitude, m	% of Forests
Plains	Over 200	13.1
Hilly Land	201-600	32.4
Lower Mountain Belt	601-1000	30.3
Middle Mountain Belt	1001-1600	21.4
High Mountain Belt	Under 1600	2.8

800m, where drought occurs most often. The distribution of precipitation in Bulgaria according to its area is also unsatisfactory. Kostov *et al.* (1976) estimated that 41.8% of the territory of the country has under 600mm precipitation, which is insufficient for normal forest production and poses the risk of drought conditions in some years.

Regional meteorological conditions, based on measurements from representative stations (the National Institute of Meteorology and Hydrology, and the Forest Research Institute, BAS) are expressed as values of de Martonne's Index of Aridity (Raev *et al.* 1995). There is a regular decrease of general humidity from west to east in Bulgaria (from 30.5 to 24.6 for North Bulgaria), related to the reduction of precipitation and increase of temperature. On average this index is 27.7 for North Bulgaria, based on one representative station, and 25.2 for South Bulgaria. Humidity conditions in eastern Bulgaria, as well as in the most southern part of the country are most unfavorable (respectively 22.5 and 22.7). However in the mountain regions of the coniferous tree belt the index is 50-65, indicating favorable humidity. The optimum zone for forest vegetation falls within 900 ± 50m to 1600 ± 100m where the aridity index is between 40 and 70 (Raev 1983).

The Impact of Drought in Representative Forest Ecosystems

To establish the impact of drought over the 1982-1994 period on forests in Bulgaria, we use data from stations of the Forest Research Institute which have been studying ecological processes in representative forest ecosystems.

Changes in Mean Annual Air Temperature

A considerable increase of mean air temperature occurred in 1990 and 1994 for oak forests (Souvorovo Station), reaching 12.0-12.9°C versus the average of 11.6°C. (Figure 10.1). The situation was similar in the transition between

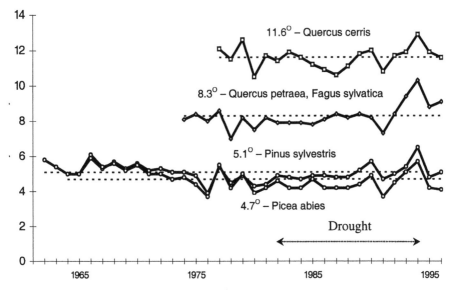

Figure 10.1 Mean annual temperature (°C) of representative forest ecosystems of Bulgaria, 1962-1996. *Quercus cerris* – Souvorovo Station; *Quercus petraea* and *Fagus sylvatica* – Gabra Station; *Pinus sylvestris* – V. Serafimov Station; *Picea abies* – Ovnarsko Station

oak and beech forests (Gabra Station) where temperatures reached 10.3°C in 1994 versus the norm of 8.3°C (Marinov 1998).

A similar maximum in representative *Pinus sylvestris* forests and *Picea abies* forests occurred in 1994, expressed respectively with 6.5°C versus an average of 5.1°C for *Pinus sylvestris* forests and 5.7°C compared to an average of 4.7°C for *Picea abies* forests.

Therefore, regardless of forest vegetation zone, the trends of change in air temperature in the main forest ecosystems in Bulgaria are similar. The highest increase of mean annual temperature for 1990 and 1994 witnessed an increase of 2°C. With the increase of altitude, this warming was lower in coniferous forests, between 1°C to 1.4°C.

Changes in Precipitation

In the zone of *Pinus sylvestris* forests, a drought period was most intense in 1993 and 1994 (Figure 10.2). Considering the average of 748mm, for the drought period, precipitation fell by 15.9% to 629.2mm, and 1993 and 1994 were respectively 47.6 and 55.1% of the norm.

Drought in Bulgaria

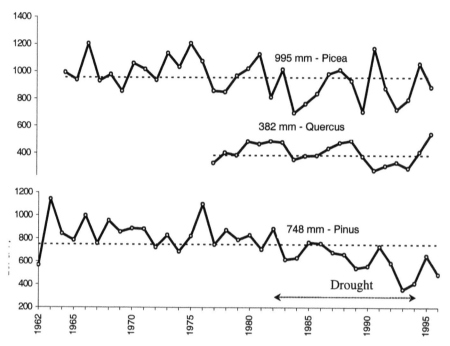

Figure 10.2 Precipitation (mm) of representative forest ecosystems of Bulgaria, 1962-1996. Top: *Picea abies –* **Ovnarsko Station;** *Quercus cerris* **– Souvorovo Station; Bottom:** *Pinus sylvestris –* **V. Serafimov Station**

To a certain extent this is valid for *Picea abies* forests. However, there were years with normal or even greater precipitation (Figure 10.2). Compared to an average of 995mm for a longer period, during 1982-1994 precipitation was 869.3mm or 87.4% of the norm. Precipitation was lowest in 1983, 1989, 1993 and 1994.

The situation in *Quercus* forests was similar. Drying was strongest throughout the 1990-1994 period, when precipitation was 84.0% of the norm (Figure 10.2). Precipitation was particularly low in 1990 and 1993, falling respectively to 71.3% and 75.4% of the norm. Throughout the 1982-1994 period precipitation was reduced in forests in Bulgaria, with minima in 1993 and 1994.

Indices of Aridity: de Martonne and Walter

Temperature and precipitation themselves are not sufficient as characteristics of drought years. The comparison of the two factors together might give a real idea of the degree of aridity. According to de Martonne's index, when values of the index exceed 40, the forest can achieve climax formation with optimal hygrothermal conditions. However, other situations can deviate from this optimum with an index from 31 to 40 during individual years; an index from 21-30, at which forests experience chronic problems with humidity and productivity falls; and an index below 20, the beginning of break-up of forest vegetation.

The mean value of de Martonne's index of aridity for *Picea abies* forests for the 1964-1997 was 65, and never reached values below 40. This means that even in drought years of 1982-1994, in *Picea abies* forests there were no problems with the hygrothermal regime. However, the values of this index over the 1982-1994 period, with the exception of 1991, varied between 40 and 20 in the zone of *Pinus sylvestris* forests. *Pinus sylvestris* forests had problems with soil moisture in 1993 and 1994, which inevitably had led to a reduction of biological productivity.

The aridity in oak forests, where the mean index for 1977-1997 was 18.5, reached values of 13-16 for the 1989-1994 period. It is a clear indication that natural *Quercus* forests in this zone were subjected to quite unfavorable hygrothermal conditions The situation in the *Fagus sylvatica* forest zone is similar to the *Picea abies* zone. Warming was strong in 1990 and 1992-1994, exceeding the 1972-1977 mean of 6.1°C, reaching up to 7.1-7.2°C. Precipitation there was minimum between 1992 and 1993. The index of aridity was largely satisfactory, except in 1975 and 1993 when aridity was below 40. It is clear from this data that beech forests had no problems with soil moisture during this period of drought.

Figure 10.3, Walter's climatogram with data from *Picea abies* forests for the 1965-1997 period, offers considerable information. The diagram proves categorically the absence of drought in *Picea abies* forests, even during the drought of the 1982-1994 period. The whole period is indicative of a good supply of moisture, which is a prerequisite for high biological activity. The situation in *Pinus sylvestris* forests is identical. However, in separate years there were conditions of brief drought periods, usually toward the end of the summer.

The scene in the lowlands, where the *Quercus* forests are situated, is different. As is evident from Figure 10.4, the annual dry period for the 1976-1997 period in northeast Bulgaria lasted from 25 June to 17 October, on average about 114 days. Thus the real vegetation period was reduced from 7 months to 3 months, which inhibited the growth of trees. According to Larher (1978), conifers require a continuous vegetation period with sufficient moisture and would find it difficult to survive under these conditions.

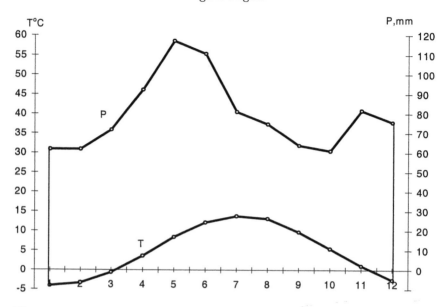

Figure 10.3 Walter Climatogram—*Picea abies* forest at Ovnarsko Station, 1965-1997

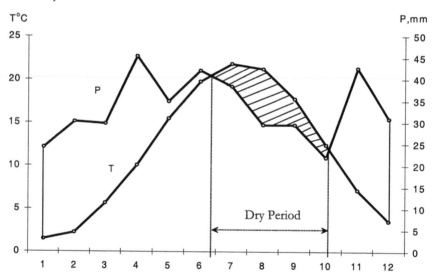

Figure 10.4 Walter Climatogram—*Quercus* forest at Souvorovo Station, 1976-1997

Moreover, this dry period affects most broadleaf species unfavorably, reducing their physiological endurance.

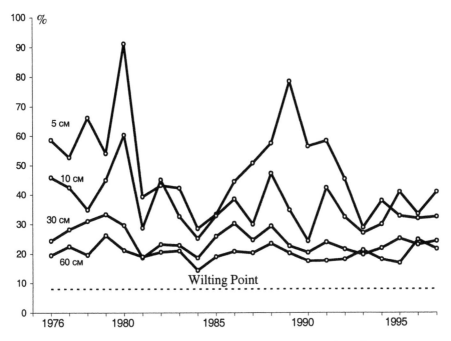

Figure 10.5 Dynamics of soil moisture in the *Picea abies* forest zone at Ovnarsko Station in the 5, 10, 30 and 60cm layers

Changes in Soil Moisture

Soil moisture provides forest vegetation with the necessary quantities of water for transpiration and synthesis of organic matter. This process is realized through the suction mechanism of the tree roots in the soil layer. Soil water falls into two categories: accessible and inaccessible. The limit of inaccessible soil moisture is determined as the permanent wilting point, which is defined as the maximum quantity of water retained against the forces of plant suction on soil particles (Rode 1969).

Soil moisture is the outcome of joint action of a number of factors: precipitation, infiltration, air temperature, quantity of evapotranspiration, including the transpiration of the plant cover. The quantity of soil moisture is an integral expression of the interaction of a number of abiotic and biotic factors and is a basic determinant of biological productivity.

Figure 10.5 is a synthesis of data for soil moisture in representative *Picea abies* forests in Bulgaria for 5cm, 10cm, 30cm and 60cm deep soil layers for the 1976-1997 period. Two main conclusions follow from the figure. Surface soil layers are subject to much greater variations in soil moisture than deep layers; and no instance occurred of inaccessible soil moisture for the 1976-1997 period, including the last dry period, i.e. *Picea abies* ecosystems had no problems concerning soil water availability.

This result categorically confirms the conclusions from other processes analyzed so far (precipitation, temperature, aridity index, Walter climatogram), namely that precipitation in the zone of natural *Picea abies* forests has not been a limiting factor of growth. *Picea abies* forests have a sufficient buffer ability to bear drought such as that of the 1982-1994 period without any particular consequences.

The situation with *Pinus sylvestris* forests (Figure 10.6) is the same, judging from the data from representative ecosystems in the V. Serafimov station for 1976-1995. In natural *Pinus sylvestris* forests only in 1993 did the value of soil moisture approach the permanent wilting point but did not cause any irreversible physiological processes. Soil moisture values were also quite low in 1994 and 1995; however, the tendency was a gradual improvement. It is evident that *Pinus sylvestris* forests in Bulgaria, similarly to *Picea abies* forests, went through the drought of 1982-1994 without any particular stress, and only had problems with soil moisture in 1993. This probably had some effect on biological productivity in the ecosystem (Raev 1998).

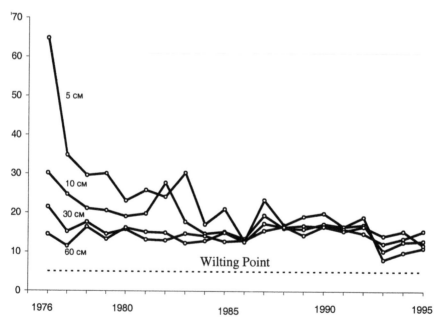

Figure 10.6 Dynamics of soil moisture in the *Pinus sylvestris* forest zone at V. Serafimov Station in the 5, 10, 30 and 60cm layers

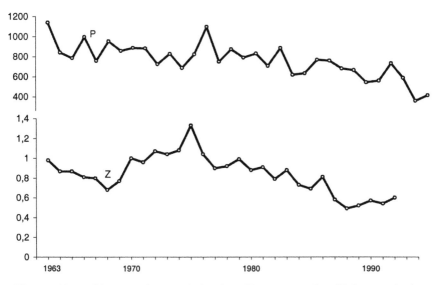

Figure 10.7 Changes in precipitation (P, mm) and radial growth (Z, mm) in *Pinus sylvestris* forest, 1963-1994

Changes in Biological Productivity

The link between biological productivity in *Pinus sylvestris* forests, expressed through the radial annual growth (in mm), and 1963-1994 precipitation dynamics are given in Figure 10.7. The data for radial growth were based on research conducted by Grozev and Delkov (1995). A common tendency in the changes can be seen as an indication of a causal link between the processes. In 1988 a minimum in radial growth was followed by a gradual increase in biological productivity. Changes in the growth vary between 0.8 and 0.5mm for the dry period. Sharp falls in productivity, as well as in the critical values of growth are absent, which is clearly related to the satisfactory hygrothermal regime in *Pinus sylvestris* forests in Bulgaria, even during the period of drought.

A very strong link between the trend of precipitation and biological productivity in forests can be seen through data from dendrochronological research in Bulgaria (Raev *et al.* 1982; Raev and Georgiev 1985; Raev *et al.* 1987; Georgiev *et al.* 1987; and others).

Drought in Bulgaria

The State of Natural Forest Ecosystems in the Drought Period

The state of natural forest ecosystems in Bulgaria is depends on the vitality of the main tree species in them—*Pinus sylvestris*, *Abies alba*, *Picea abies*, of the conifers, and *Fagus sylvatica*, *Quercus sp.* of the broadleaf species—which occupy approximately 70% of the forest area. Their distribution is well organized according to zones and belts, *Quercus sp.* being the main species in the lower belt at an altitude up to 700-800m, *Fagus sylvatica*, *Abies alba* and *Pinus sylvestris* in the medium belt from 700m to 1200-1300m, and *Picea abies*, *Pinus peuce*, *Pinus heldreichi* in the belt above 1300m. This distribution in zones clearly shows that the impact of drought on forest vegetation was different (expressed through climatic parameters; Raev 1998).

Assessment of the state of forests in this study is a summary and analysis of data obtained between 1982 to 1995 through research projects (Donov *et al.* 1982; Rosnev *et al.* 1993; Alexandrov and Rosnev 1992; Rosnev 1994; Rosnev and Bobev 1994; Rosnev and Petkov 1999; and others), annual reports of forest pathology studies of the Forest Protection Stations at the National Forest Service (NFS), NFS bulletins (1990-1995), and statistical data from the Committee for Forests up to 1990-1995.

The dynamics of damage for different species, their foliage, branches and crowns, including the type of damage and the pests that cause it, have been traced on representative stands for separate habitats gathered through the Monitoring of Forest Ecosystems (NU/ECU 1988-1995).

Table 10.4 shows the loss and coloration of leaves used in ecological monitoring of forests in Europe. A complex evaluation of the state of health and vitality of separate trees and stands is given in Table 10.5. Damage from diseases, and pests which caused a decline in the state of the trees and stands have also been assessed.

The state of natural conifers is comparatively good. Conifers account for about 30.3% of the total area of forests in Bulgaria. *Pinus sylvestris* stands cover about 165,000ha; *Pinus nigra*, 29,000ha; *Abies alba*, 31,000ha; *Picea abies*, 150,000ha; and other conifers, a total of 38 000ha.

Table 10.4 Scale for leaf loss and coloration

Grade	Leaf-loss, %	Coloration, %	Damage
0	0 – 10	0 – 10	Healthy
1	11 – 25	11 – 25	Slight damage
2	26 – 60	26 – 60	Average damage
3	under 60	under 60	Heavy damage
4	100	100	Dry

Table 10.5   Scale for complex assessment

Leaf-loss (degrees)	Coloration (grade)			
	0	1	2	3
	Complex assessment (degrees)*			
0	0	0	1	2
1	1	1	2	2
2	2	2	3	3
3	3	3	3	3
4	4	4	4	4

*Complex assessment (degrees): **0** – healthy, **1** - slight damage, **2** - average damage, **3** - heavy damage, **4** – dry.

Natural coniferous ecosystems (without *Abies alba*) are quite stable. Partial drying and damage has occurred largely due to abiotic factors (felling by high winds, snow cover) and a limited number of biotic factors, representing between 0.49-0.68% of their area (Table 10.6). Separate instances of drying were established for the 1993-1995 period, related to attacks on *Picea abies* by *Ips typographus L.* and periodic diseases of the leaves caused by *Lophodermium macrosporum (Hort.) Rehm* and *Chrosomixa abietis (Walle.) Ung.* Damage from *Heterobasidion annosum (Fr.) Bref.* is quite serious, however the hidden manner of development of the pathogen (in the core of the stem) chiefly affects its growth without causing full die-back of trees on a massive scale. Natural *Pinus sylvestris, Pinus nigra,* as well as *Pinus heldreihi* and *Pinus peuce* stands are in a very good state.

Abies alba Arn. was badly affected by drying in the regions of the western and central Rhodope mountains, northern Pirin and eastern Rila. *Abies alba* is a very sensitive tree species to climatic anomalies as well as air pollution. Drying occurs chiefly in stands up to 1000-1200m at an age between 80-100 years. The data in Table 10.6 shows that the area of drying stands (over 20% of the trees) was 12.98% for 1989, and those with over 50%, 0-0.54%. The high intensity of drying continued until 1995. The course of the process is shown on Figure 10.8.

The marked drying of *Abies alba* over the 1982-1995 period led to an invasion of *Pityocteines carvidens (Germar)* and *Pityocteines spinidens (Reiffer)*, which considerably accelerate the drying process and increase the loss of timber and biomass. The data show that the continuous drought after 1982 had an accumulative effect on the vitality of tree species, which reacted in a marked manner in 4-5 years after the beginning of the dry period. This is confirmed by the results of observations over representative *Abies alba* stands at an altitude of 1300 and 1500m. In the first case, drying losses are much greater,

Table 10.6 Forest area affected by drying and damage, 1988–1995 (natural ecosystems)

| Type of forest | Total area (ha) | Area affected by drying in % | | | | | |
| | | 1989 | | 1993 | | 1995 | |
		over 20%	over 50%	over 20%	over 50%	over 20%	over 50%
Natural Conifer without *Abies alba*	376,000	-	-	0.68	0.00	0.49	0,00
Abies alba	31,000	12.98	0.54	7.21	0.22	11.06	3.06
Oaks	791,900	3.77	0.19	6.83	0.34	5.42	0.44
Quercus cerris	258,400	0.72	0.09	21.38	1.56	15.50	1.62
Fagus sp.	446,600	0.19	0.00	1.19	0.00	0.94	0.00

while the second drought was barely felt (Ikrishte locality, North Pirin) and no drying process occurred.

Studies of *Fagus sylvatica* and *Quercus sp.* provide an idea of the state of natural broadleaf forests. The data in Table 10.6 show a comparatively good state of beech forests, growing at altitudes 700-800 to 1200m. Only separate instances of drying and group drying occurred over 0.19% to 1.19% of their area of 446,600ha, chiefly in eastern Stara Planina and the southern slopes of Sredna Stara Planina. The table shows that drying increased in 1993 and appeared as a development of the disease along the branches, caused by the fungus *Nectria ditissima Tul* and leaf necrosis by the pest *Rhynchaenus fagi* L. The state of beech forests is assessed as comparatively good.

The health and vitality of oak forests is considerably worse. Studies by Rosnev and Zlatanov (1983), Rosnev and Alexandrov (1992), Rosnev (1993),

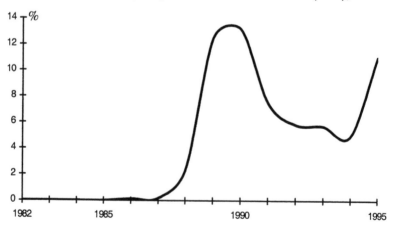

Figure 10.8 Dynamics of drying of the natural *Abies alba* forests

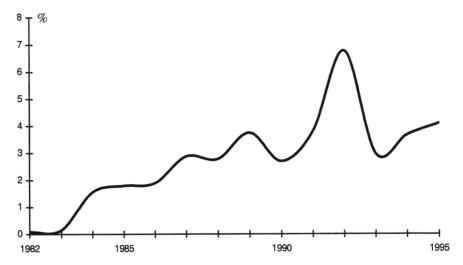

Figure 10.9   Dynamics of drying of the natural *Quercus* forests

and Rosnev and Bobev (1993), together with data from studies on forest pathology, carried out by the Forest Protection Stations (1982-1995) show that the area of oak forests affected by this drought throughout this period has constantly been growing (Figure 10.9) from 0.15% (1240ha) in 1983 to 6.78% (54,120ha) in 1991. What is characteristic is that drying initially affected only *Quercus petrae*, where the subsequent damage was greatest (the area includes stands with losses over 20%). Foci of drying occur in oak stands, which is a negative factor for the stands. The strongest impact is in forests of *Quercus petraea*, the least with *Quercus frainetto* and *Quercus pubescens (Willd)*. Damage came from pests attacking the stem, woodworm (*Ips typographicus*) and as well as tracheomycotic fungi *Ophiostoma sp.* and woodworms of the genera *Armillaria, Polyporus, Driadeus* etc. Serious damage has been caused by leaf-eating insects *Limantria dispar L.*, the species of the genera *Geometridae* and *Buprescidae*. According to Tsankov *et al.* (1997) between 56,000ha and 207,000ha of oak forests have been infested by leaf-eating insects with a maximum during 1983-1984 and 1992-1995. The fight with these pests is chiefly through biological measures. Biological damage is part of the set of factors affecting the health of oak forests.

A considerable part of oak forests consists of *Quercus cerris L.*, a species considered resistant to drought and biotic damage. Initially its state of health was better than that of *Quercus petraea*. Up to 1989 damage and drying on a mass scale were limited (from 0.09 to 0.19% of their total area, 258,400ha). After that period, in the beginning in northeastern Bulgaria and later in the remaining part of the country, drought began on a massive scale, and considerable damage occurred from *Hypoxilon mediteraneum (de Not) Mill,* together with *Diplodia mutila Fr. Mont.* In 1993 areas affected by drying of stands (over a range of 20 to 50% of the trees) reached 21.38%, while those above 50% were 1.56%. The high intensity of the process continued after remedial steps up to 1995. (Table 10.6). The conclusions of these studies (Rosnev and Petkov 1999) show continuous drought and the chronic impact of biotic factors, with fungi and insects as the main cause of the observed physiological disturbances. The management of these forests and their age structure are also a factor.

The data cited and the dynamics of pathological processes in natural forests in Bulgaria bring out certain regularities between climatic change and the health and vitality of the main tree species. These results clearly outline the leading role of climate on the functioning of ecosystems, most clearly expressed in the lower oak forest belt.

Drought During 1982-1994 and Climate Change

Analysis of the data on climatological impacts during the drought period on forests in Bulgaria clearly shows that the influences of drought are not identical for the different forest vegetation zones in Bulgaria.

Warming in the zone of natural coniferous forests and beech varied between 1.0°C to 1.4°C; precipitation reduction was from 12.0% to 15.9%; de Martonne's index of aridity was over 40 for *Picea abies* and *Fagus sylvatica*, while that for *Pinus sylvestris* was between 40 and 20. Under these hygrothermic conditions, natural *Picea abies* and *Fagus sylvatica* forests in Bulgaria have had no problems with soil moisture needs, while certain difficulties existed only in some isolated instances for a short period. This conclusion was also confirmed by Walter's climatograms together with data on soil moisture.

Nevertheless, in lower lying lands under 800m, the altitude of oaks, warming reached 2°C mean annual temperature. Precipitation was lower on average by 24.9% to 28.7% of the norm, while the index of aridity reached the critical values of 13-16. Under these conditions the period of drought continued annually without interruption from the 25th of June to October 17th (114 days). In this way conditions were extremely unfavorable for natural oak vegetation and other species as well, which explains the drying of natural forests of *Quercus petraea, Quercus cerris* and other species.

These drought parameters over the 1982-1994 period fall within the "optimistic" scenarios of climate change over the coming 50 or 100 years, as projected in the beginning and mid-1990s (IPCC 1992; ANL 1994), together with basic scenarios of possible climatic change throughout the 21st century, created in the late 1990s (Flato *et al.* 1999; IPCC-DDC 1999 *et al.*).

It is clear that we have the rare chance to reflect the real picture of future change in forest climate and register the reaction of forest vegetation in terms of hygrothermic stress conditions. These conclusions could assist in the compiling of an action plan for the forestry sector and taking measures for the adaptation of forests in Bulgaria in view of possible climatic change.

Conclusions

Throughout the 1982-1994 period the increase of air temperature was highest in 1990 and 1994, reaching an increase in mean annual temperature by 2°C in the zone of broadleaf forests. With an increase of altitude, warming in natural coniferous forests varies between 1°C up to 1.4°C.

Over the study period, a clearly expressed depression in precipitation was established in all forest zones in Bulgaria. It was lowest in vegetation at the higher altitudes of forest vegetation (about 12% below the norm); in *Pinus sylvestris* forests it was about 15.9%, in *Quercus sp.* forests reaching over separate years a reduction of 24.9-28.7%. The fall in precipitation was most serious in 1993 and 1994.

No problems appeared in humidity (index of de Martonne above 40) in natural *Picea abies* and *Pinus sp.* forests, situated in the highest parts of the country throughout this period of drought. In *Pinus sylvestris* forests a certain disruption occurred in the water regime throughout this period (index between 40 and 20). However, in the zone of oak forests during the period a chronic deficit of humidity occurred (index of aridity 13 and 25), which inevitably was reflected in the physiological processes of forest vegetation.

Similar conclusions also appear in the analysis of Walter's climatograms: natural *Picea abies* forests had no problems with humidity; *Pinus sylvestris* forests had certain problems only in separate years, and in these instances it was over a short period. The situation with *Fagus sylvatica* forests is identical. However a continuous drought period (over 100 days) occurred in *Quercus* sp. forests, which led to a real reduction of the vegetation period by half with a sharp fall in hygrothermal conditions in this part of the country.

The data for soil moisture in natural *Picea abies* forests clearly shows that even throughout the drought period there was no shortage of soil moisture. The situation with *Fagus sylvatica* forests was identical. In the zone of *Pinus sylvestris* forests soil moisture reached this limit only in specific years, but did not exceed it.

A strong link exists between biological productivity of tree vegetation determined according to the dendrochronological method and the annual course of precipitation, in particular in the low-lying forest vegetation zone.

Natural *Pinus sylvestris*, *Abies alba*, *Fagus sylvatica*, *Picea abies* and *Pinus nigra* forests during the drought period showed no substantial negative changes. *Abies alba* forests, however, which are most sensitive to drought, suffered considerable damage. The situation with *Quercus petraea* forests, as well as *Quercus cerris*, situated in the low-lying vegetation zone, where the index of aridity was quite low, is similar. On the other hand drought had covered the greater part of the vegetation period. Physiological disturbances in tree growth caused by drought are a pre-condition for greater damage from pests and diseases, which accelerate the processes of damage and drought.

Studies clearly show that over the 1982-1994 period a level of climatic change had set in corresponding to the main scenario of possible climatic change in the 21st century. Through this period of crisis, stress caused by climatic changes in the zone of natural *Fagus sylvatica* and coniferous forests in Bulgaria was slight and did not led to serious losses. *Abies alba* forests, which are highly sensitive and suffered considerable losses, are an exception. The consequences for natural forest vegetation in the low-lying lands, with their more intensive drought, were more serious, leading to great damage for *Quercus petraea* and *Quercus cerris* forests, as well as to other tree species.

References

Alexandrov A. and B. Rosnev. 1992. "Oak Decline in Bulgaria," *Forest Science* 1:3-6.

Argonne National Laboratory (ANL). 1994. *Guidance for Vulnerability and Adaptation Assessments.* U. S. Country Studies Program, Washington D.C.

de Martonne, E. 1925. *Traite de Geographie Physique*, Paris: Colin.

Donov, V. *et al.* 1991 Експертна оценка за здравословното състояние на горските култури и някои горски насаждения в България.

Flato, G. M. *et al.* 1999. "The Canadian Centre for Climate Modelling and Analysis, Global Coupled Model and Its Climate," *Climate Dynamics* 16(6):451-467.

Georgiev, N., I. Raev and O. Grozev. 1987. "Tree-ring Investigations on *Picea abies* Forests in the Western Rhodope Mountains for the Purposes of Ecological Monitoring," pp. 30-35 in *Dendrochronological Methods in Forest Science and Ecological Forecasting.* Irkutsk.

Grozev, O. and A. Delkov. 1995. Дендрохронологична оценка за състоянието на бялборовите екосистеми в биосферния резерват "Дупката", Западни Родопи. *Сп. Наука за гората*, кн. 2, pp. 20-27.

Intergovernmental Panel on Climate Change (IPCC). 1992. *Climate Change: The IPCC 1990 and 1992 Assessments.* WMO, UNEP.

Intergovernmental Panel on Climate Change Data Distribution Centre (IPCC-DDC). 1999. Data information supplied by the IPPC Data Distribution Centre of Climate Change and Related Scenarios for Impact Assessments, Version 1.0, April 1999.

Kostov, P., I. Zhelev, P. Belyakov, S. Angelov, K. Kaloudin, K. Tashkov, R. Rashkov, P. Nasalevska. 1976. Състояние на горските ресурси на НРБ и съображения за тяхното комплексно използуване. Замиздат, София.

Larher, V. 1978. Экология растений. Moscow: Mir.

Marinov, I. T. 1998. Екологичен стационар "Габра." Информация за валежи и температура. София: Отчет на БАН.

National Forest Board. 1982-1995. Годишни отчети за състоянието на основните екосистеми в България на ЛЗС - София, Пловдив, Варна при НУГ за периода 1982-1995г., МЗГ.

Raev, I, N. Geogiev, E. Dimitrov, H. Peev. 1982. Опит за дендрохронологичен анализ на гори от Quercus ceris в Девненския промишлен район. *Горскостопанска наука* 3:3-15.

Raev, I. and N. Georgiev. 1985. Многогодишни вариации в растежа на гори от Quercus ceris в Странджа планина. Горскостопанска наука 5:3-13.

Raev, I., N. Georgiev, L. Rousseva, B. Baikov. 1987. "Tree-ring Investigations of *Picea abies* Forests in Parangalitsa Biosphere Reserve," pp. 195-206 in *Methods of Dendrochronology-I, Proc. Methodology of Dendrochronology: East-West Approaches, 2-6 June, 1986, Krakow, Poland,* Warsaw.

Raev, I. 1989. Изследвания върху хидрологичната роля на иглолистните горски екосистеми в България. Автореферат на докторска дисертация, Author's summary of Ph.D. dissertation, Bulgarian Academy of Sciences.

Raev, I. 1993. Опит за диференциране на планинските климати в Рила и зоната за оптимално горскостопанско производство, *Proceedings International Symposium on the Man-Mountain Ecosystems Relationships,* Project 6, MAB, Vratsa, 24-29 October 1983, vol. II, 227-236.

Raev, I. 1994. Някои резултати от многогодишните стационарни изследвания в горите на България, провеждани от Института за гората при БАН. В: Национална конференция по лесозащита и мониторинг на горските екосистеми, ИГ, МОС, КГ; София, 30-31.03. 1994 г., стр. 149-182.

Raev, I., O. Grosev and V. Alexandrov. 1995. Проблемът за бъдещите климатични промени и противоерозионните залесявания в България. В: Национална конференция с международно участие "90 години борба с ерозията на почвата в България", 16-20 октомври 1995 г., София, стр. 84-93.

Raev, I., O. Grozev, V. Alexandrov, Z. Vassilev, B. Rosnev, A. Delkov. 1996. Оценка потенциала на горите за поглъщане на парникови газове. Уязвимост спрямо климатичните промени и мерки за адаптация на горската и земеделската растителност в България. Смекчаване влиянието на климата в неенергийния сектор. 2.1. Сектор "Гори." В: Изследване за България насочено към глобалните проблеми на изменение на климата, Енергопроект - ЕАД, София.

Raev, I. 1997. Климатична характеристика и динамика през периода 1986-1996г. В: Екомониторинг на горите в Рила (1986-1995), ОМ2, кн. 5:18-26.

Raev, I. 1998. "The Hydrological Role of Forests—A Key Ecological Factor," pp. 53-64 in *Congress of Ecologists of the Republic of Macedonia with International Participation, 20-24 Sept. 1998,* Ohrid.

Rode, A. A. 1969. Основы учения о почвеной влаге. Т. II, Гидрометеоиздат, Ленинград.

Rosnev, B. and S. Zlatanov. 1982. Трахеомикозно заболяване на зимния дъб в Източна Стара планина. Горско стопанство, кн. 8:50-54.

Rosnev, B. and P. Petkov, 1992. Върху патологични причини за влошаване на здравословното състояние на цера (Quercus ceris L.) в България. Сб. Национална конференция по лесозащита, Sofia, стр. 45-52.

Rosnev, B. 1993. Здравословно състояние на културите от бял, черен бор и дугласка в България. *Бюлетин по лесозащита* 3:22-31.

Rosnev, B. 1994. Здравословно състояние на дъбовите гори в България през периода 1980-1992г. Сб. Доклади на Националната конференция по проблемите на издънковите гори, Sofia, стр. 71-77.

Rosnev, B. and R. Bobev. 1994. Изменения в здравословното състояние на дъбовите гори през последните години, Доклади на конференция на РДГ - Бургас, ИГ-БАН.

Rosnev, B. *et al.* 1994. Експертна оценка на здравословното състояние на белия и черния бор в България. Отчет по договор с Комитет по горите, ИГ-БАН.

Rosnev, B. and P. Petkov. 1997. Здравословно състояние на белия бор (Pinus silvestris L.) в представителни насаждения на Рила планина, *Наука за гората* 3-4:66-70.

Rosnev, B. *et al.* 1997. Оценка и мониторинг на въздействието на вредните вещества във въздуха върху горските екосистеми. Отчет по програмите NU/ECE за периода 1986-1995г.

Tsankov, G., P. Mirchev and G. Georgiev. 1997. Видов състав и структура на вредната листогризеща ентомофауна в дъбовите гори на България. *Acta Entomologica Bulgarica*, 1-2:66-69.

Walter, H. 1973. *Vegetation of the Earth in Relation to Climate and the Eco-physiological Conditions.* New York: Springer Verlag.

Chapter 11

Population Dynamics of Birds and Mammals During the Drought

Georgi Markov, Milena Gospodinova, Grigor Penev

The temporal and spatial variation of physical environmental conditions has both direct and indirect influence on animals. Climate is the main factor of the physical environment that significantly affects the biota in ecosystems through the temperature regime and the presence of water. Variation of climate conditions is often one of the factors that regulate the population of different animal species in natural ecosystems (Whittaker 1975; Soule 1978).

Comparison of the population dynamics of species during normal and drought conditions would help to outline the possible influence of changed drought conditions on the development of wild animals (Odum 1971; Gaughley 1977). One must take into account that this analysis is obtained without assessment of many other factors, such as the general state of environment (natural succession), anthropogenic effects, predator effects, and illegal hunting, which also affect the wild animal population. The absence of scientific data prevented us from estimating these factors.

The assumptions that all the individuals react identically to climatic conditions typical for the drought period as well as the elimination of some factors influencing the wild animal population simplify the model of animal-environment interaction. In this way, using existing official data, the first and the only picture of development of wild animal population in natural Bulgarian ecosystems during drought was created.

There are many approaches to estimate the population dynamics of wild animals (Moss *et al.* 1982). The most applicable approach was to analyze available data from of wild animal counts. The population of some of them had to be determined indirectly, for example, relative to the number of individuals shot during the hunting season (Grigorov 1982) because of their hidden way of life and/or missing counts for a long enough period of time.

Comparative analysis of the changes in population dynamics of numerous species with different ecological niches increases the reliability of estimation of the influence of climate. The correct interpretation of the observed dynamics changes requires good knowledge of ecological and biological features of the species, in which population dynamics is investigated at extreme changes in the climate conditions.

One of the crucial factors for the development of mammalian populations is the presence of clear water (Markov 1959). Climatic conditions, especially changes in the precipitation and the presence of enough water, have different influences on each species. This necessitates that changes in the population of species inhabiting different ecological niches be analyzed concurrently, such as by parallel population study of wild boar (*Sus scrofa*), European brown hare (*Lepus europeus*), chamois *(Rupicapra rupicapra)* and the red fox (*Vulpes vulpes*). The wild boar is widely spread in the country and needs large quantities of water. The brown hare is also wide spread and its development depends highly on the presence of an abundant nutritional base, closely related to microclimatic conditions. The chamois is a grazing ruminant inhabiting very specific rocky, inaccessible, mountainous habitats. The small rodents represent one of the components of the nutritional base of the red fox, and they are also directly affected by climatic conditions.

In order to extend the species complex and to obtain more precise interpretation of the impact of part of the drought period (1982-1991) on wild fauna, birds have also been included in this analysis. Four species were chosen: the pheasant (*Phasianus colchicus*), the capercaillie grouse (*Tetrao urogalius*), the partridge (*Predix predix*) and the rock partridge *(Alectoris graeca graeca)*. Their population in natural ecosystems is influenced by climatic factors (temperature and rainfall) both indirectly through the available food and directly through the reduction of population especially during the nesting period (Dudzinski 1977).

Research Methods

The population of the four mammal species (European brown hare, red fox, chamois and wild boar) and four bird species (pheasant, partridge, rock partridge and capercaillie), inhabiting different ecological niches was analyzed for the period 1961-1991. This period included 20 years (1961-1981) characterized by typical climatic conditions and from 1982 to 1991, which covers a part of the drought period. The representative data for the red fox are for the period 1961-1987. The temporal variation of the population dynamics for European brown hare and fox was estimated on the basis of the registered number of individuals shot per year. In the case of chamois, wild boar, pheasant, partridge, rock partridge and capercaillie these characteristics were assessed according to the official results of spring game counts. All data have been taken from the register of the National Forestry Office of the

Ministry of Agriculture and Forests of the Republic of Bulgaria. The change in population (Q) of the studied animal species is determined by calculating the difference in its population between two consecutive years (Odum 1971), $Q = Q_n - Q_{n-1}$, where Q_n and Q_{n-1} are the population in two consecutive years. Q is expressed in percentage in the figures in this chapter (see Appendix 11.1).

Results and Discussion

A generalized picture of the population of studied species during two periods, before the drought (1961-1981) when the climatic conditions were typical for their habitats and during a part of the drought (1982-1991), was created. Because of the possible biases in determination of the absolute values of the population, the main conclusions about trends were based on a comparative analysis of the differences in population between every two consecutive years.

The analysis revealed that the population dynamics of the wild boar (Figure 11.1) did not follow a specific course during the drought period. There were successive years with declining and increasing population density of this species during the period before the drought and during the drought in Bulgaria. Chamois population density fluctuated less during the drought period and although at reduced level, except in 1986, it maintained a positive trend (Figure 11.1). The European brown hare displayed population changes close to zero (Figure 11.1). It was characterized by sharp decline at the beginning and at the end of the drought period followed by growth in the subsequent year.

The overall trend in the population development of the red fox during the drought period was negative with wide fluctuations in its population (Figure 11.2). Pheasant population dynamics during the drought period exhibited entirely negative changes, although the values varied yearly (Figure 11.2). The population of the rock partridge showed small changes (Figure 11.2), slightly negative for the most years. This pattern of development during the drought period was in contrast to the interval preceding the drought, where high amplitudes of the fluctuations were usual.

The development of the rock partridge population (Figure 11.3) in the drought period was quite similar, characterized by insignificant variations around zero. The course of the capercaillie (grouse) population dynamics did not differ (Figure 11.3).

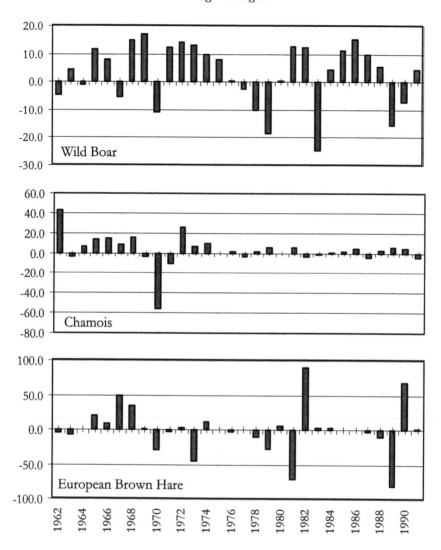

Figure 11.1 Percent change in the number of wild boar, chamois, and European brown hare from the previous year, 1962-1991. Top and middle: numbers counted; bottom: numbers shot.

Conclusions

The population dynamics found in species with different ecological niches is of exceptional importance for the correct assessment of the influence of

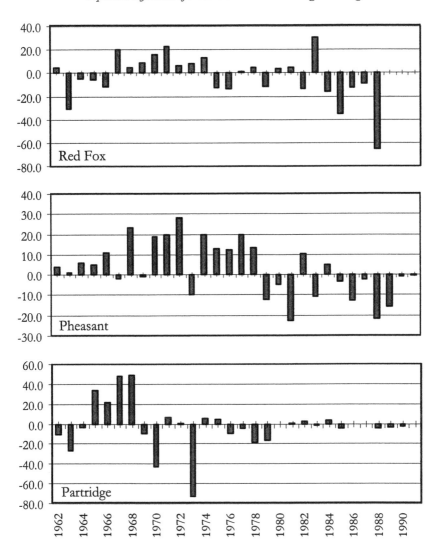

Figure 11.2 Percent change in the number of red fox, pheasant, and partridge from the previous year, 1962-1991. Top: numbers shot; middle and bottom: numbers counted.

drought on the development of wild fauna. The results of this analysis revealed a general trend in population dynamics of the studied species during the drought period. This change was more strongly expressed in wild birds than in the mammals.

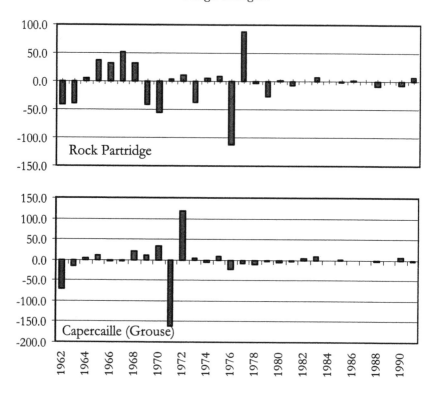

Figure 11.3 Percent change in the number of rock partridge and capercaille from the previous year, 1962-1991, based on numbers counted

The change in wild animal populations in natural ecosystems is a complex function of numerous endo- and exogenic factors, so the variation found should not be considered as determined only by changed climatic conditions. The influence of the ecological conditions is always refracted through the quantity and quality of the populations. To establish the effect of each change in the main ecological components is a complicated task, requiring sustained ecological investigations.

Animal species inhabiting the territory of Bulgaria are well adapted to normal variations in humidity. If there are no disastrous changes, the humidity has indirect effect on the wild fauna mostly through its influence on the vegetation. Precipitation and humidity determine vegetative development and wealth, and hence the presence of food and shelter for wild animals.

Recommendations for Prevention of Harm from Drought

Despite the fact that the relationship between the change in climatic conditions in the wild animals' habitats and the change in their population is quite relative, helping the maintenance of populations has to be carried out when unfavorable climatic conditions occur. Aid in natural feeding is quite important for decreased losses of wild animals and especially of game caused by drought. Improvement of game food sources in the forest includes appropriate forest management, reclamation of forest glades and pastures, and creation of game fields. The natural feeding of the game in the field must be supported by lasting changes in the landscape, resulting in formation of game fields and brushwood in the areas unsuitable for agriculture—ravines, desolated plots near water basins, gullies, etc. The presence or absence of water basins in natural ecosystems is a determinant for certain birds and mammals. The enlargement of water areas during drought could be done through building of ponds or water supplies in game reserves (Botev 1981).

References

Botev, N. 1981. Ловно стопанство. Sofia. Zemızdat.

Dudzınski, W. 1977. Patki towne. Panstwowe wydawnıctwo rolnize i lesne. Warsaw.

Gaughley, G. 1977. *Analysis of Vertebrate Populations.* London. John Wiley and Sons.

Grigirov, G. 1982. Критичен преглед на някои мнения относно причините за намаляване на численоста на дивия заек (Lepus europeus Pall.) в България през периода 1969-1979 г. *Gorskostopanska nauka,* 5:63-71.

Markov, G. 1959. Бозайниците в България. Sofia. Nauka i Izkustvo.

Moss, R., A. Watson and J. Ollason. 1982. *Animal Population Dynamics.* London. Chapman & Hall.

Odum, E. 1971. *Fundaments of Ecology.* Philadelphıa: W.B. Sunders Company.

Soule, M. (editor). 1978. *Viable Populations for Conservation.* Cambridge: Cambrige University Press.

Whittaker, R. 1975. *Communities and Ecosystems.* Second Editıon. New York. Macmillian.

Appendix 11.1

Numbers of wild boar (1), chamois (2), pheasant (5), partridge (6), rock partridge (7) and capercaillie (grouse, 8) according to spring counts and the numbers of hare (3) and red fox (4) shot by calendar year

Year	1	2	3	4	5	6	7	8
1961	12693	444	699366	59028	7634	934970	202310	4106
1962	11220	1065	664220	61690	18935	845633	147317	2425
1963	12621	1018	593370	42279	22026	616729	94329	2051
1964	12295	1116	588212	39128	38309	590879	101012	2135
1965	15956	1315	766535	35721	51296	888135	153089	2386
1966	18456	1531	841945	28205	80422	1082278	197562	2291
1967	16801	1660	1287837	40695	75131	1499513	269211	2238
1968	21377	1890	1604185	43558	137966	1928387	311865	2765
1969	26713	1843	1616761	48931	134981	1841416	257060	3011
1970	23338	1028	1358078	58517	185351	1464480	182032	3828
1971	27160	870	1328760	72744	238490	1525470	186690	n/a
1972	31602	1244	1352400	76796	314500	1534200	200400	2844
1973	35700	1346	942000	82053	288000	898000	150000	2934
1974	38758	1485	1043000	90518	342000	946000	157600	2828
1975	41228	1481	1040190	82474	376850	986170	168840	3050
1976	41312	1500	1012620	74206	410010	906060	15750	2500
1977	40500	1450	1014070	75168	463000	866750	134730	2300
1978	37400	1480	920800	77820	498800	708500	129300	2035
1979	31704	1558	669733	70711	465214	564040	91900	1944
1980	31850	1553	725050	73121	452195	561100	95160	1822
1981	35816	1632	81164	75800	391324	561930	85682	1756
1982	39665	1581	890300	67332	419000	580400	85700	1831
1983	32095	1564	914160	86108	389640	571970	96190	2068
1984	33485	1576	933960	76260	403780	600255	96120	2037
1985	37045	1601	928950	54566	394700	560220	93050	2104
1986	41774	1670	930290	46705	359840	555590	95510	2068
1987	44825	1610	901800	40895	353745	553585	96270	2042
1988	46483	1650	810990	n/a	295210	514150	85180	1946
1989	41688	1731	73750	n/a	252920	485350	84590	1906
1990	39467	1801	691843	n/a	250218	464129	74342	2089
1991	40832	1742	700537	n/a	248665	459478	84498	2010

PART VI
DROUGHT IMPACTS ON MANAGED ECOSYSTEMS

Chapter 12

Drought Impacts on Managed Forests

Ivan Raev, Boyan Rosnev

The condition of artificial forest ecosystems, planted during large scale afforestation over the last 40-50 years in Bulgaria, is of great interest to foresters. Analysis of their condition during and after the 1982-1994 dry period is particularly important. Approximately 1,200,000ha forests were created over the 1955-1995 period in Bulgaria. About 1/3 of forests in Bulgaria were replanted with new stands, and 80-90% of these young stands were conifers, mainly, *Pinus sylvestris* (Scots or white pine), *Pinus nigra* (Austrian or black pine), and *Picea abies* (northern spruce). The majority of these stands are in the low lying forest vegetation belt, below 800-1000m elevation. In the last 10 years massive drying and dieback of coniferous stands were observed.

Unfortunately, precise information on perished coniferous stands is not available. Thus, according to the statistics of the National Forestry Administration at the Ministry of Agriculture and Forests, in 1955 coniferous forests accounted only for 518,000ha (14% of the forest area). In 1990 after intensive afforestation, the number rose to 1,395,000ha or 37% of the forest area. In 1995, however, for the first time coniferous forests started to decrease in area, remaining at 1,232,000ha (32.7% of the forest area). The most important cause for this decline was the massive dieback in the 1990-1994 period, at the height of the greatest drought. Approximately 162,110ha or 18.5% of the new coniferous stands died as a result of the drought by 1995. It should be mentioned that there was no indication of damaged stands in the optimal zone of coniferous trees, in the 900-1600m elevation range (Raev 1983). Almost all dry trees were in the low forest zone, out of their natural range. Data about damaged artificial young deciduous stands in the low forest area are not available when stands consist of local species.

Causes for the Drying of Coniferous Stands

What is the cause of the drying of young coniferous monoculture stands in the low forest area? We should look for the answer to this question first in the difference between the water balance of the introduced coniferous versus local deciduous species which are well adapted to the drought conditions.

This difference was assessed through the use of direct water balance measurements for two species quite widespread in Bulgaria—*Quercus cerris*, which occupies a large part of the low forest area, and *Pinus nigra*, which is the most popular coniferous species in Bulgaria, introduced in the low forest area over the last decades (Table 12.1). The experiment was carried out in the Souvorovo Forestry Commission in northeastern Bulgaria; measurements were taken in 1983 as part of a survey of the dry 1982-1994 period.

The main difference in the water balance of the two species is in the amount of rain water retained within and evaporated from tree crowns, technically known as *interception*. While interception of *Quercus cerris* is only 41.3mm (11.7% of the precipitation, which was 325.7mm in 1983), that of *Pinus nigra* reached 107mm, or 2.6 times more (32.9% of the precipitation). Thus for transpiration, i.e. biomass accumulation, *Pinus nigra* had only 139.6mm of transpiration, which is insufficient for sustaining high vitality and productivity. At the same time we should remember that this moisture was spent as early as the second vegetative month (May). In the June-September period all physiological processes virtually stop, creating stressful conditions for *Pinus nigra*. Apart from reduced productivity, drought resistance was weakened sharply, and the species easily fell victim to secondary diseases and pests.

With *Quercus cerris*, 216.2mm transpiration is almost sufficient for normal vegetative growth, so *Quercus sp.* survive comparatively well at low precipitation values, considering that the structure of fascicular bundles in

Table 12.1 *Quercus cerris* and *Pinus nigra* water balance at Souvorovo Forestry Commission, Northeast Bulgaria, 1983

Water balance elements	Index	*Quercus cerris*		*Pinus nigra*	
		mm	%	mm	%
Precipitation to the crowns	P	352.7	100.0	352.7	100.0
Precipitation below the crowns	P'	291.5	82.6	236.3	66.9
Stem flow	F_{st}	19.9	5.6	9.4	2.7
Precipitation reaching the soil	P''	311.4	88.2	245.7	69.6
Interception	E_i	41.3	11.7	107.0	30.3
Soil evaporation	E_s	95.2	27.0	106.1	30.1
Transpiration	E_t	216.2	61.3	139.6	39.6
Total evaporation	E	352.7	100.0	352.7	100.0
Infiltration	F_g	0.0	0.0	0.0	0.0

deciduous species is better adapted for quicker passage of water from the root system. In a short time *Quercus cerris* accumulates normal annual biomass. In contrast, coniferous trees because of their narrower channels have a slower water exchange, and require a constant water income over the course of many months. Hence, dry periods inevitably lead to water stress (Larher 1978). The defoliation season in deciduous species (October-March), when almost all total precipitation reaches the soil and creates water reserves, contributes to low interception losses (Raev 1989). It is clear that deciduous species are better adapted to insufficient precipitation and should be preferred for afforestation at low elevations (Raev 1977, 1995, 1997; Raev *et al.* 1994).

In extremely dry years in the low elevation part of northeastern Bulgaria, such as the 1990 precipitation of 272.2mm and 1993 of 288.1mm, the evaporation component of the water balance for forest vegetation consisted only of interception and soil evaporation (136-215mm) annually. In practicality, this means that there was no water left for productive transpiration. Under these conditions a crisis situation arises even for such drought-resistant species as *Quercus cerris* (in the low lying areas) and *Quercus petraea* (in the hilly parts of the country).

Artificial Forest Ecosystem Health During Drought Periods

In 1990 artificial forest ecosystems filled approximately 900,000ha of the forested land, 685,000 of which were *Pinus sylvestris* and *Pinus nigra* stands, and 170-190,000 were deciduous species (poplars, acacias, birches, walnut, oaks).

Coniferous species–*Pinus sylvestris, P. nigra, Pseudotsuga mensiesi, Abies alba* and *Picea abies*–form the basic part of Bulgarian artificial forests. More than 90% of them are *Pinus sylvestris* and *P. nigra* planted on waste land and reconstructed low-value oak and hornbeam (*Carpinus betulus*) areas up to 600-700m above sea level. The health of these ecosystems is shown in Table 12.2.

In 1986 the area of coniferous species stands where more than 20% were affected by drought was 5299ha (0.8%). In 1989 it rose to 24,250ha (3.51%), and in 1995 it reached 37,390ha (5.44%) of the forest area. The number of severely affected forests with over 50% of dried trees was also on the rise from 1167ha (0.17%) to 5150ha (0.75%) in 1993. Figure 12.2 presents the course of the process. The dieback of *Pinus* stands started in 1984 and was followed by a fast increase, reaching two peaks, 1988-1990, and a higher one in 1992-1995. Dieback of *Pinus sylvestris* and *Pinus nigra* develops in one and the same way (Table 12.2), but with *Pinus nigra* in areas of low altitude, it is more intense.

Table 12.2 Area of artificial coniferous forest ecosystems affected by drought

Type of forest	Total area, ha	Area of damaged by drought, %					
		1989		1993		1995	
		>20%	>50%	>20%	>50%	>20%	>50%
Planted coniferous forests	687,300	3.51	0.17	5.44	0.75	2.87	0.57
P. silvestris	401,800	2.01	0.13	4.26	0.65	2.68	0.63
P. nigra	285,500	5.60	0.21	7.10	0.88	3.15	0.49

The complex assessment of forest health done according to international methods is very symptomatic of the condition of individual trees in representative sample areas in different regions of Bulgaria. This assessment combines defoliation (needle shedding) and leaf coloring data. It also uses a rating score of the degree of damage (Chapter 10, Table 10.5, this volume).

The results of the survey for the 1988-1995 period shown in Figures 12.1, 12.2, and 12.3 display the scope and dynamics of leaf damage in *Pinus nigra* and *Pinus sylvestris* stands. Evidently, *Pinus nigra* trees in the eastern part of the Stara Planina mountain and southwestern Bulgaria with more than 26% needle damage in their crown stood out with their intensity. Fast growth caused considerable damage by the *Sphaeropsis sapinea* fungus, which seriously affects *Pinus nigra* vegetation, is typical of the former region.

Pinus sylvestris defoliation and leaf coloring almost completely coincides with the forest pathology survey; more serious damage occurred in 1988-1991 and after 1993-1994. On *Pinus sylvestris,* fast growth of *Lophodermium* and *Gremmeniella* fungi occurred. In addition, in some regions there was serious

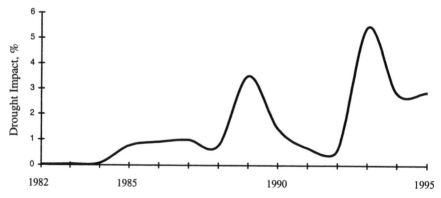

Figure 12.1 Area of artificial coniferous forests affected by drought

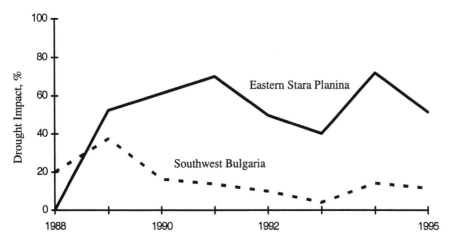

Figure 12.2 Comparative data for affected *Pinus nigra* forests (2-4 degrees of damage)

damage caused by *Heterobasidion annosum*, a root fungus causing local drying in artificial ecosystems.

An analysis of the changes in ecological factors and their impact on coniferous forests points to the severe drought after 1982 as the main factor. We also should bear in mind that during large scale afforestation in the low lying mountain belt, the adaptability of pines to climate variability was

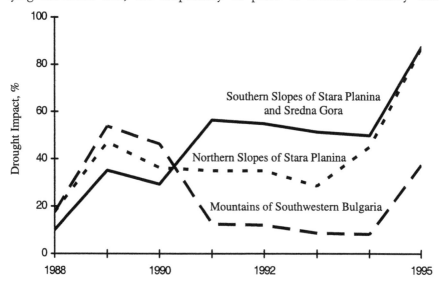

Figure 12.3 Comparative data for affected *Pinus sylvestris* forests (2-4 degrees of damage)

overestimated. The physiological weakening of the trees is a major prerequisite for the rapid development of diseases and insect pests. In most of the cases they damage the whole tree system, which speeds up drying and dying.

The condition of deciduous artificial ecosystems has deteriorated over the last years in spite of the fact that they are fewer in number. Collected data on acacia, red oak, walnut and poplars illustrate this phenomenon. For example, acacias afforested in perfectly suitable areas were affected by climate changes, with two peaks of drying, 1986-1989 and after 1992. With red oak stands, particularly grave drying was recorded in the Byala Slatina Station, where there is a continental climate. In some areas the number of drought-affected trees varied between 20% and 100%, while in other parts of the country similar damage was insignificant (Rosnev and Petkov 1994).

The general conclusion is that artificial forest ecosystems are more seriously affected by drought than natural ones, which fact is a reflection of the continuous drought and chronic air pollution in some regions. Diseases, pests and uninformed afforestation policy are also important factors.

Forest Fire Dynamics in the Forest Fund of Bulgaria during Drought

Figure 12.4 traces the change in the number of fires and the area of burned forests in Bulgaria over the 1984-1997 period according to the statistics of the National Forest Administration at the Ministry of Agriculture and Forests. The average number of forest fires varied between 50 and 150 fires covering 500-2000ha per year. These numbers indicate that for decades on end, Bulgaria was not a country for which the problem was typical (1960-1990).

Even during the first drought stage, for example in the 1985-1989 period, an average of 102 fires per year totaling 620ha is an insignificant percentage of Bulgaria's forest area (3,400,000ha). The average annual number of fires for these five years was 549, an increase of 5.4 times. Much more drastic is the increase of burned forest area: from 602ha in the preceding period to 7633ha annually or 12.7 times in the analyzed period. In 1993 and 1994 there was a forest fire maximum. In 1993 there were 1196 fires over 17,264ha, which amounts to 11.7 times more fires and 28.7 times larger areas than in normal years. The area burned in 1993 was twice that of new annual afforestation in recent years. Obviously this is a clear illustration of ecological catastrophe in Bulgarian forests.

The number of forest fires coincides with the severest drought in 1993 and 1994. No doubt, the major factor for the sky-rocketing forest fires in the

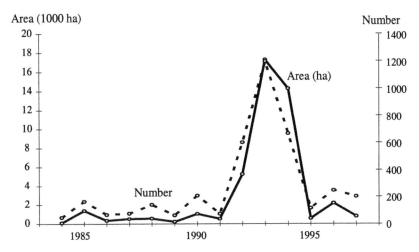

Figure 12.4 Number and area of forest fires in Bulgaria

1990s was the extensive warming combined with less precipitation, particularly in the low forest area. However, there is considerable evidence that some of the fires were caused by arson, provoked by the restitution of forests and agricultural lands. Moreover, statistical data show that over two-thirds of the burned forests were conifer stands, planted out of their natural habitat, i.e. below 800m elevation. This is another example of the wrong practice of coniferous afforestation in the low regions of the country, in the climax habitats of oak, beech, hornbeam and other species (Raev 1977, 1989).

Forest Vulnerability and Adaptation to Climate Change

The Bulgarian participation in the U.S. Country Studies Program included a section on conditions and vulnerability of forest vegetation to future climate change (Raev *et al.* 1995, 1996; Grozev *et al.* 1996; Raev and Grozev 1999; etc.). Central to the research was a program for increasing forest adaptation to future climate warming and aridification. The results were included in the *First* and *Second National Communications on Climate Change* (Republic of Bulgaria 1996, 1998) presented as official documents under Bulgaria's obligations to the Framework Convention on Climate Change (1992). This program is also included in the *National Climate Change Action Plan* (1999).

The measures proposed in the *Action Plan* should become the guidelines in the *Strategy for the Development of Forestry* in Bulgaria for the next decades, and they could limit the negative effects of future climate warming and drought. What are the main points in this program?

Low lying forests (below 800m elevation). On the threshold of possible climatic change, the strategic task for forestry in this most vulnerable part of forests in Bulgaria should be to adapt forests to climate aridification and preserve forests from deteriorated ecological conditions using local thermophilic tree and shrub species and naturalized foreign species and species from neighboring climatic zones with proven endurance in semi-arid conditions. Secondary forests should be converted into seed forests by introducing more enduring and thermophilic species. Seedling production should provide for the gradual replacement of existing local and foreign tree and shrub species to tree and shrub species more adapted to air pollution. New soil preparation, afforestation and secondary afforestation techniques should be devised to meet the change in ecological conditions and species. A differentiated approach to forestry in different regions of the country must recognize that aridity is expected to intensify in a spatial way: western plains-North Bulgaria-South Bulgaria-the Black Sea region-Southwest.

Forests in higher parts of the country (over 800m elevation). Thanks to the possibilities for maintaining more favorable conditions in the higher regions of Bulgaria, forestry should aim at the preservation of the biodiversity in forests, sustainable development of ecosystems, and multifunctual use and development of the national system of protected natural territories. The mixed character of mountainous forest ecosystems by extensive use of the available deciduous and coniferous stock should be preserved. Thus it will be possible to meet the increasing demand for th enrichment of forest biodiversity. The lower border of forests should be raised by some 100-200m in connection with the climate changes in Bulgarian mountains, with new forest and shrub species introduced to enrich biodiversity and increase ecosystem productivity and resistance. The moratorium on clear cuts should remain in effect to preserve the most valuable high-stem forests. Water catchment area management should lead to multi-purpose use: maximal biological production combined with the best water regulation, soil protection, recreation, and microclimatic diversity. All-round development of protected areas, national parks, reserves, protected forests, etc., will preserve biodiversity and maintain high productive capacity and the protective potential of forests.

In all forest areas, preservation and extension of the mixed character of forests should be preserved, giving preference to all climate-change-resistant tree and shrub species. In all more favorable conditions, multi-level ecosystems in which trees are of different age should be created. Better logging methods for thinning should be developed in line with the new conditions and species, and intensive sanitary felling should be used for the immediate removal of diseased trees. An adequate system for prevention, immediate localization and extinguishing of forest fires is imperative, as is development of a program for fighting pests and diseases in forest and shrub vegetation during deteriorated climate conditions. A system for forest protection from trespass and criminal action should be developed and implemented. The implementation of a program for increasing forest adaptability to climate changes is the road to sustainable development and management of Bulgarian forestry in times of ecological stress (Raev 1999).

References

Grozev, O., V. Alexandrov and I. Raev. 1996. "Vulnerability and Adaptation Assessment of Forest Vegetation in Bulgaria," pp. 374-383 in L. Smith *et al.* (editors), *Adapting to Climatic Change, Assessment and Issues*. New York: Springer-Verlag.

Larher, V. 1978. Экология растений. Moscow, Mir.

National Climate Change Action Plan. 1999. Bulgarian Ministry of the Environment, Energoproekt, Sofia.

Raev, I. 1977. Върху хидрологичния ефект на насаждения от по-главните наши дървесни видове. *Горско стопанство* 7:1-7.

Raev, I. 1983. Опит за диференциране на планинските климати в Рила и зоната за оптимално горскостопанско производство. *Proceedings of the International Symposium, The Man-Mountainous Ecosystems Relationship, Project 6 – MAB*, Vratsa, 24-29 October 1983, кн. 2, стр. 227-236.

Raev, I. 1989. Изследвания върху хидрологичната роля на иглолистните горски екосистеми в България. Author's summary of Ph.D. dissertation, Sofia, Bulgarian Academy of Sciences.

Raev, I., G. Antonov, G. Popov, P. Mirchev, P. Petkov. 1991. Причини за съхненето на иглолистните култури в Североизточна България. *Горско стопанство* 1:19-21.

Raev, I. 1994. Някои резултати от многогодишните стационарни изследвания в горите на България, провеждани от Института за гората при БАН, pp.149-182 in *National Forestry and Ecosystems Monitoring Conference, the Forest Institute, Ministry of Environment, Forestry Committee;* Sofia.

Raev, I. 1995. Главни причини за съхненето на иглолистните горски култури в България. *Гора* 1:25-28.

Raev, I., O. Grosev and V. Alexandrov. 1995. Проблемът за бъдещите климатични промени и противоерозионните залесявания в България. В: Национална конференция с международно участие "90 години борба с ерозията на почвата в България", 16-20 октомври 1995 г., София, стр. 84-93.

Raev, I., O. Grozev, V. Alexandrov, Z. Vassilev, B. Rosnev, A. Delkov. 1996. Оценка потенциала на горите за поглъщане на парникови газове. Уязвимост спрямо климатичните промени и мерки за адаптация на горската и земеделската растителност в България. Смекчаване влиянието на климата в неенергийния сектор. *2.1. Section: Forests in Bulgarian Research for the Global Problems of Climate Change,* Sofia: Energoproect EAD (PLC).

Raev, I. 1997. Климатична характеристика и динамика през периода 1986-1996 г., pp. 18-26 in *Ecomonitoring on Forests in the Rila Mountain (1986-1995),* OM2, Book 5.

Raev, I. 1999. "Sustainable Forestry Management," pp. 217-229 in B. Nath, L. Hens, P. Compton and D. Devuyst (editors), *Environmental Management in Practice, 2. Compartments, Stessors and Sectors,* London and New York: Routledge and UNESCO.

Raev, I. and O. Grozev. 1999. Горската покривка, биопродуктивността и промените в климата, Христов Т. и др., Глобалните промени в България. София, Национален координационен център по глобални промени, стр. 173-194.

Republic of Bulgaria. 1996. *The First National Communication on Climate Change.* Sofia: Ministry of Environment and Water, Pensoft.

Republic of Bulgaria. 1998. *The Second National Communication on Climate Change.* Sofia: Ministry of Environment and Water.

Rosnev B. and P. Petkov. 1992. Върху патологични причини за влошаване на здравословното състояние на цера (*Quercus cerris* L.) в България, pp. 45-52 in Miscellany, *National Forest Protection Conference,* Sofia.

Rosnev, B. 1993. Здравословно състояние на културите от бял, черен бор и дугласка в България. *Forest Protection Bulletin* 3:22-31.

Rosnev B. and P. Petkov. 1997. Здравословно състояние на белия бор (*Pinus Sylvestris* L.) в представителни насаждения на Рила планина, *Наука за гората* 3-4:66-70.

Chapter 13

Drought Impacts on Crops

Vesselin Alexandrov, Nicola Slavov

Drought is considered to be one of the greatest natural disasters. In comparison to all other natural disasters, droughts occur most frequently, have the longest duration, cover the largest area, and cause the greatest loss to agricultural production. In Bulgaria, severe droughts have become a major climatic disaster in recent decades, adversely affecting the country's needed agricultural products and water supplies.

Influence of Meteorological Variations on Maize Grain Yield

Unfavorable meteorological conditions cause twice as many losses in agricultural production as they do in most other business sectors. The intensity, time of occurrence, duration and area of impact are important dimensions of unfavorable meteorological events when they seriously decrease crop yields. Drought is usually the most important constraint limiting crop production in the rain-fed areas in the country. Drought may be considered a severe meteorological event, for example, when it covers more than 10% of the sowed arable land, when atmospheric humidity is less than 30%, or when less than ten millimeters of soil moisture is in the top soil layer for ten or more days.

The soil and climatic conditions in Bulgaria are suitable for cultivation of maize. However, when precipitation deficits and high air temperatures occur during the critical periods of maize development, the crop's growth, development, and yield formation are limited.

The influence of variations of air temperature and, especially, precipitation on the maize grain yield in Bulgaria has been analyzed (Alexandrov 1995; Alexandrov and Slavov 1994, 1998; Slavov and Alexandrov 1994a,b,c; Tomov *et al.* 1996b). Maize yield data from 21 experimental crop stations across Bulgaria were used in the current study. The Bulgarian maize hybrid "Kneja 611" and winter wheat variety "Sadovo 1" were planted each year from 1970 to 1993. The data were obtained from the

State Crop Variety Commission (SCVC). Daily weather data, including precipitation and air temperature, were assembled for the same period at the nearest neighbor weather stations.

Figure 13.1 shows the long-term fluctuations (1970-1993) and trends of mean air temperature and precipitation across the Bulgarian territory during the periods April to September for elevations below 800m. Yearly grain yields are also presented. During the period 1970-1993, the trends of these meteorological variables diverged; air temperature increased whereas precipitation decreased. Variations of air temperature and especially of precipitation are highly correlated to national maize production. A decrease in maize productivity in 1970, 1974, 1985, 1990 and 1993 was observed due to a precipitation deficit during the crop growing season (Figure 13.1). The decreasing trend in maize production was observed in all non-irrigated and in most irrigated crop experimental stations (Figure 13.2).

The deviation from normal summer precipitation since 1970 has influenced maize yield. Simple linear regression analyses suggest that maize yield in Bulgaria is mainly affected by precipitation in July and August (Table 13.1 and Figure 13.3). Most of the statistical models included precipitation during the warmest month of the year, e.g., July. Precipitation in August was also important because drought conditions usually occur during this month. Soil moisture in April is important to document because planting usually occurs between April 10 and April 30. This is partly why precipitation in March has a positive impact on final maize yield. However, maize yields in regard to precipitation trends are mediated by temperature; higher temperatures in April and in July can affect maize yield.

Statistical models often provide the basis for estimating regional production for a given set of climatic conditions. Therefore, an attempt was made to develop statistical models for regional assessment of maize productivity in Bulgaria. For the northern part of the country and the non-irrigated crop experiment stations across Bulgaria, equations included average precipitation in March, July and August, with correlation coefficients of 0.73 and 0.73 respectively. Simulated grain yield of maize in North Bulgaria was in accordance with the measured data for the calibration data (1970-1990) as well as the evaluation data (1991-1993). Differences near 20% between calculated and measured grain yield were observed for only 1974 and 1985, when a significant drought occurred.

Precipitation during some of the critical phenological phases was found to correlate well to final grain yield. Relationships developed for the Kneja Station using results from 16 field experiments conducted with unchanged agrotechnology during the period 1967-1970 included precipitation during the

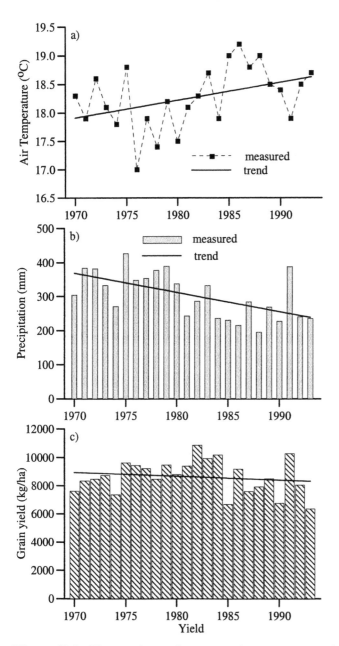

Figure 13.1 Fluctuations of average air temperature (a) and precipitation (b) during April-September, and maize yield averaged for crop experiment stations in Bulgaria (c)

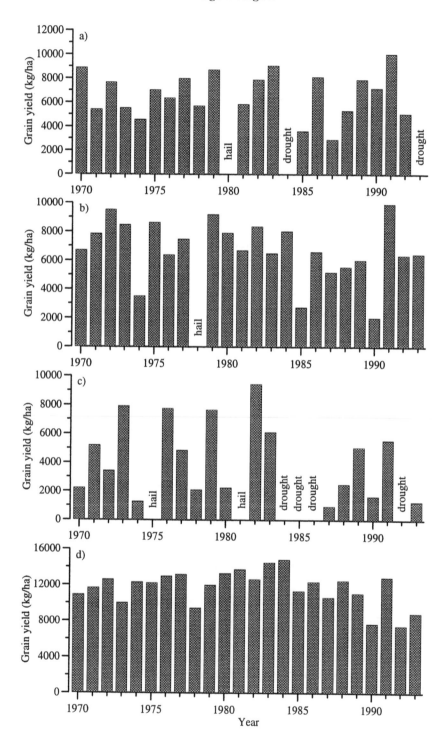

Table 13.1 Correlation between maize grain yield and precipitation during different time intervals, averaged for non-irrigated crop experiment stations in Bulgaria (1970-1993)

Period	July-Sept.	Aug.-Sept.	August	July	April-June	June-July
Correlation Coefficient	0.59*	0.37	0.46*	0.62*	0.14	0.57*
Period	April-August	April-Sept.	April-July	May-August	June-August	July-August
Correlation Coefficient	0.53*	0.50*	0.47*	0.51*	0.63*	0.64*

* 5% significance level

period between leaf formation and tasselling and from tasselling to silking as significant variables ($r = 0.89$). Figure 13.4 shows the comparison between measured and calculated grain yield for this station.

Precipitation during the period from leaf formation to tasselling of maize, plant density, and total amount of fertilizer applied were included in a second statistical analysis for the same station developed using data from 48 field experiments during the period 1967-1974 ($r = 0.91$).

Using selection and genetic materials, it is possible to create maize lines and hybrids more tolerant to drought conditions. An analysis of the data representing various inbred maize lines showed significant variability in tolerance and vulnerability to drought stress and high temperatures (Tomov 1989; Tomov *et al.* 1996a). The most significant finding was that the most stable maize lines under drought were lines created from Bulgarian selection materials. Through observed variability from stable to vulnerable maize lines on a large scale, it may be possible to hypothesize that the crop tolerance to drought depends on several complex genetic factors. Similar crop reactions to drought stress can be also observed in the case of using world inbred maize lines. It is interesting to mention that the reactions of the same inbred maize lines were different under different treatments. It is clear that the conditions for maize cultivation were different. The obtained results suggest that classic inbred maize lines are highly heterogeneous in their reaction to drought stress despite being homogeneous according to base indicators.

Figure 13.2 (opposite) Fluctuations of maize grain yield in three non-irrigated stations (a – Medkovets, b – Tsarev Brod, c – Zimnitsa) and one irrigated station (d – Gorski Izvor)

Drought in Bulgaria

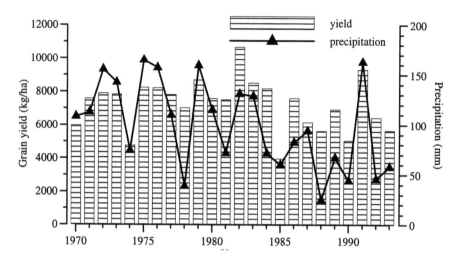

Figure 13.3   Fluctuations of maize grain yield and precipitation in July and August, averaged for non-irrigated stations in Bulgaria

Drought Impacts on Crop Production During the 1990s

A large deficiency in precipitation was observed during the July through September period of 1992 (Slavov and Alexandrov 1993), making it one of the driest Bulgarian summers of the century. Drought conditions occurred mainly in August and September, two months that normally mark the peak of the dry season in Bulgaria. Precipitation through the first ten and twenty days was below 0.2mm. Precipitation in the third 10 days in August and first ten days in September was considerably below 100-year averages. No precipitation above 1mm occurred during the rest of September. Precipitation alone, however, does not give an accurate indication of drought. The 1992 drought was also combined with high air temperatures (32-37°C), making available soil moisture in the top meter about 50-55% of capacity in August and September, which is insufficient for crop growth. River flows in the country were about 30% below the average values.

The 1992 drought persisted through the autumn and winter seasons. Available soil moisture in the top 1m of soil was 55%, 65%, and 60% of its available capacity in the months of October, November, and December, respectively. The 1992-1993 winter dryness caused extremely low soil moisture nationwide; the observed soil moisture conditions were the driest measured since 1970. Mountain snow packs for December 1992-January 1993

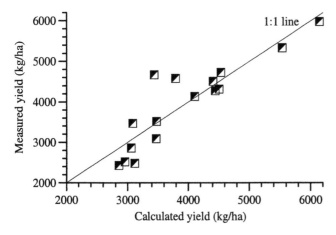

Figure 13.4   Comparison between measured and calculated grain maize yield at Station Kneja (1967-1970)

were considerably below average for this time of the year, and the water reservoirs generally did not contain an adequate water supply.

Monthly rankings of precipitation across the country show that January 1993 was unusually dry in most areas; snowfall above 1mm occurred only on the second and third days of the month, and snow cover was poor. January was also characterized by abnormally warm conditions, helping to make available soil moisture in the top 1m 70-75% of its available water holding capacity.

In February, Bulgaria received less than half of its average precipitation. Available soil moisture still remained in a critical deficit situation. Bulgaria's soil moisture conditions in January and February 1993 were the driest recorded since 1970. During March, however, the precipitation pattern changed. Precipitation in this month was about normal. Nevertheless, drought conditions persisted in March because of the precipitation deficit from August 1992 to February 1993. The winter precipitation total for 1993 (50mm or about 50% of normal) was one of the lowest in the last one hundred years. Below-normal precipitation during the recharge period, November 1992-March 1993, caused generally low soil moisture reserves in the spring of 1993.

Water supplies increased until 20 April 1993, and then abruptly returned to normal. At the end of April, the soil moisture in the upper layer (0-20cm) was 10-25mm, which is insufficient for normal emergence and growth of spring crops. The development of winter crops was also delayed. In some areas of the country the available soil moisture in the top meter was only 60-70% of the available capacity in April. In May, precipitation was nearly normal and agrometeorological conditions were favorable across the country. In June, however, national precipitation was 50% of the mean. Available soil moisture during June in the top 1m was 40-70%, which was insufficient for normal growth of spring crops, fruit trees and vegetables. The root zone of crops in this month was very dry. To make matters worse, the last ten days of June were characterized by abnormally high air temperatures.

The drought conditions persisted and worsened in July: monthly precipitation was below 30% of the mean during this month. The upper layer (0-20cm) contained little or no soil moisture, and available soil moisture in the top 1m was 35-50% of the available capacity. Crop growth was poor: winter crops were harvested, but the maize crop did not begin tassel initiation, and it dried or stayed undersized. Sunflower anthesis extended too long without insemination. Roots like sugar-beets had slow biomass formation, bean crops dried without pod formation, and grapes remained small.

The drought expanded in August, with continuous precipitation deficit and high air temperatures, which slowed the growth of most crops. Below-normal precipitation during this month led to untimely maturity of spring crops and lower yields. Much of the maize crop for grain production was used for silage. Drought conditions persisted during September and October with national precipitation between 30 and 40% of the mean. Available soil moisture was insufficient for normal germination and emergence of the sowed winter crops in October (Slavov and Alexandrov 1993, 1994b).

The summer drought of 1993 adversely affected Bulgaria's agricultural sector. Crop losses caused mainly by drought in that year are presented in Table 13.2. The data were provided by the National Statistical Institute in Sofia. One can easily assess the damages provoked by weather whims.

An assessment of drought tolerance and the influence of the genotype on maize growth and final yield was conducted in 1993. A good productive potential of national (e.g., Kn-613 and Kn-614) and foreign maize hybrids, which are included in the base production of maize in Bulgaria, was outlined despite significant yield variability. The new hybrids of maize production in the country showed their productive potential: Kn-613, Kn-614, Kn-704, Russe 464, RK 588. Similar results were obtained using the foreign maize hybrids Pioneer 3737 and Volga (Tomov *et al.* 1995, 1996b).

An investigation of the development and dynamic reaction to the drought stress in 1993 was also done using a visual assessment and a detailed analysis of crop changes and the formation of reproductive organs (Tomov *et al.* 1989, 1996a). Some maize hybrids reacted negatively and quickly stopped the physiological functions of the leaves (see Appendix 13.1). As a result, the maize productivity changed and the number of harvested maize ears and grain

Table 13.2 Crop losses caused by drought in Bulgaria, 1993

Crop	Mean yield 1989-1992 [kg/ha]	Yield 1993 [kg/ha]	Sowed areas 1993 [10^3 ha]	Decrease in yield [10^3 t]	Decrease in production [10^3 t]	Loss [10^6 US$]
Wheat	4050	2870	1266	1494	747	72.3
Barley	3800	2650	361	415	207	16.7
Maize	3700	2270	370	529	958	102.2
Sunflower	1580	940	403	258	116	14.6
Sugar Beets	21170	11000	10	104	42	0.6
Cotton	960	740	7	16	6	18.0
Tobacco	1240	1130	32	35	12	21.7
Fruits						9.7
Wine Grapes	4600	3500	105	116	29	2.8
Dess. Grapes	5200	4520	12	8	2	0.3
Total						258.9

yield decreased. Grain yield varied widely between 907 and 3355 kg/ha. The unfavorable meteorological conditions in 1993 created a drought stress for the maize crop, cultivated mainly in West Bulgaria. As a result of the drought in this region, maize completely dried and was used for silage. In spite of the drought conditions in the western part of the country in 1993, a relatively high production of hybrids Kn-530, Kn-611, Kn-613 and Kn-614 was observed in the rest of the Bulgarian territory.

Measures for Crop Adaptation to Drought During the 21st Century

The expected climate change conditions in Bulgaria were estimated at the National Institute of Meteorology and Hydrology using outputs of global circulation models (GCMs) such as GFDL, GISS, CCC, OSU, UKMO, and others. Most of the GCMs used in the studies projected air temperature increases and precipitation reduction (especially in the summer) during the 21st century in Southeast Europe and Bulgaria (Alexandrov 1997a, 1997b, 1999a). The potential impact of global warming on agricultural crop growth, development, and yield formation was also assessed (Alexandrov 1997a, 1997b, 1999a, 1999b, 2000; Slavov and Alexandrov 1996a, 1996b, 1997). Under the current level of CO_2, the GCM climate change scenarios projected a decrease of winter wheat and especially maize yield during the 21st century. Yield decreases were mainly expected to be caused by a shorter crop growing season and precipitation deficits during the warm-half of the year. When the direct effects of CO_2 were included in the study, all GCM scenarios resulted in an increase in winter wheat yield. These possible results can affect the forage balance of the country in the future.

An "Action Plan" in Bulgarian agriculture was developed by the Ministry of Agriculture and Forests. The results obtained during the investigation of the climate change impacts in the country forced the Action Plan to develop two major goals: to decrease greenhouse gas emissions to the atmosphere and to adapt agriculture to climate change.

The expected climate change in Bulgaria will require new crop zoning in several directions, including: (a) Increasing areas of the most important agricultural crops over new regions that are characterized by improved thermal and moisture conditions; (b) Utilization of more varied cultivars and hybrids, especially long-maturing, high-productive cultivars and hybrids with better industrial qualities; and (c) Cultivation of new agricultural crops grown in Mediterranean areas. New cultivars and hybrids will be adapted to climate change with the following goals in mind: (a) The new cultivars of winter agricultural crops should pass the winter season of organogenesis under higher temperatures without departures in the normal crop growth and development; (b) The new cultivars and hybrids should be drought-resistant cultivars and hybrids, especially at the end of the vegetative period and at the beginning of the reproductive period; (c) Higher maximal air temperatures should not provoke thermal stress effects, especially during crop flowering and formation of the reproductive organs; and (d) The new cultivars and hybrids should take advantage of an increased concentration of atmospheric carbon dioxide.

The changes in climate during the 21st century will cause significant changes in agrotechnological management. Measures related to all agro-technological management and practices under climate change were included in the Action Plan, paying special attention to limited water resources. The following measures for increasing irrigation effectiveness should be applied: (a) Introduction of irrigation technologies with decreased water charges and without losses to water transport and distribution; (b) Restoration and reconstruction of the previously constructed irrigation facilities; (c) Reconstruction and building of new test-pits for utilization of groundwater; (d) Utilization of river water and precipitation for storing irrigation water during the winter season; and (e) Re-utilization of waste water and drainage system water.

Future irrigation technology should be oriented to the development and utilization of irrigation systems which economically use water resources, losing less water during water transportation and irrigation. There are several projects already underway in the country that are assessing the necessity and effectiveness of systems for drip irrigation.

Water losses through seepage and evaporation in canal and flood irrigation systems can be minimized by lining the canals with cement or switching to pipe irrigation systems. Irrigation should be applied only early in the morning and during the late evenings in order to keep evaporation at a minimum. The irrigation rates should not be high, about 400-500m³/ha under strong water control. Significantly higher costs of production related to irrigation systems will most likely be offset by shifts of water use to areas where there are lower rates of moisture loss.

Changes in the types of agricultural production and irrigation systems will require significant changes in farm layout and the types of equipment employed. In areas that require irrigation systems, additional water reservoirs or boreholes may be needed. In reaction to recent droughts, the government has embarked on the construction of a number of medium to large-sized dams throughout the country. It is thought that because of the large costs involved in infrastructural changes (at the farm level) only small incremental adjustments may occur without changes in government policy.

On the other hand, appropriate technologies for collection and storage of snow-melt water should be used due to the expected increase of precipitation during the cold half of the year, as has been shown by most GCM simulations. Precipitation during the cold half the year can also be used for irrigation of trees and winter crops using relevant technologies and systems.

Research institutes and universities in Bulgaria are working on solutions to the irrigation problems. Numerical irrigation experiments using the computerized Decision Support System for Agrotechnology Transfer (DSSAT) are carried out at the National Institute of Meteorology and Hydrology to simulate the optimum water amounts for irrigation under climate change scenarios (Alexandrov 1998, 1999a; Slavov and Alexandrov 1998b). The simulations are carried out with respect to biophysical and economic analyses of the harvested maize yield and net returns.

Using groundwater for irrigation in Bulgaria is a problem which is still not well investigated. Through 1990, groundwater was used for irrigation in only 4% of the total irrigated land in the country. This practice was not appropriately used in Bulgaria during its drought years in the 1990s. It is important to note that one of the important research priorities of the European Commission is the utilization and management of groundwater. This priority should not be avoided in Bulgaria because a precipitation deficit during the warm-half of the year is projected. It should also be noted that

there exists one more alternative for irrigation in agriculture in Bulgaria—
utilization of waste water (Petrov 1994, 1995, 1997). Further comprehensive
investigation is needed.

References

Alexandrov, V. 1995. "Climate Variability and Drought in Bulgaria," pp.35-42 in N. Tsiourtis (editor), *Water Resources Management under Drought or Water Shortage Conditions.* Rotterdam: Balkema.

Alexandrov, V. 1997a. "Vulnerability of Agronomic Systems in Bulgaria," *Climatic Change* 36: 135-149.

Alexandrov, V. 1997b. "GCM Climate Change Scenarios for Bulgaria," *Bulgarian Journal of Meteorology and Hydrology* 4(4):205-211.

Alexandrov, V. 1998. "A Strategy Evaluation of Irrigation Management of Maize Crop under Climate Change in Bulgaria," Vol.3, pp.1545-1555 in *Proceedings of the Second International Conference on Climate and Water*, Espoo, Finland.

Alexandrov, V. 1999a. "Vulnerability and Adaptation of Agronomic Systems in Bulgaria," *Climate Research*, 12(2-3):161-173.

Alexandrov, V. 1999b. Земеделието и промените в климата, Христов, Т. и др., Глобалните промени в България, София: БАН, стр.195-218.

Alexandrov, V. and G. Hoogenboom. 2000. "The Impact of Climate Variability and Change on Crop Yield in Bulgaria," *Agricultural and Forest Meteorology*, 104(4):315-327.

Alexandrov, V. and N. Slavov. 1998. Колебания на добива на царевица в зависимост от метеорологичните условия, Растениевъдни науки, т. 35, № 1, стр.11-17.

Petrov, K. 1994. "Irrigation with Water Containing Heavy Metals," *Proceedings of the 17th European Regional Conference on Irrigation and Drainage*, ICID, CIID, Vol.3, pp.294-298.

Petrov, K. 1995. "The Optimal Control of the Irrigation Systems under Drought or Water Shortage Conditions," pp.273-276 in N. Tsiourtis (editor), *Water Resources Management under Drought or Water Shortage Conditions.* Rotterdam: Balkema.

Petrov, K. 1997. "Environmental and Demand Management of Irrigation under Water Shortage Conditions in Bulgaria," pp.9-14 in J. Refsgaard and E.Karalis (editors), *Operational Water Management.* Rotterdam, Balkema.

Slavov, N. and V. Alexandrov. 1993. "Drought in Bulgaria, Update and Historical Perspective," *Drought Network News* 5(2):12-15.

Slavov, N. and V. Alexandrov, 1994a. "Drought and its Impact on Grain Yield of Maize," Vol. 1, pp. 321 – 327 in *17-th European Regional Conference on Irrigation and Drainage*.

Slavov, N. and V. Alexandrov. 1994b. "Persistent Drought in 1993 Affects Bulgarian Agriculture," *Drought Network News*, 6(1):19-20.

Slavov, N. and V. Alexandrov. 1994c. Влиянието на колебанията на метеорологичните условия върху добива на зърно от царевицата в България, Сб. V-ти нац. симпозиум "Физика и селско стопанство", стр.224-229.

Slavov, N. and V. Alexandrov. 1996a. Влияние на бъдещото изменение на климата върху агроклиматичните ресурси на България, Сп. Растениевъдни науки. т.XXXII кн.9, стр.72-77.

Slavov, N. and V. Alexandrov. 1996b. Уязвимост спрямо климатичните промени и мерки за адаптация на земеделската растителност в България, Сб. Изследване за България насочено към глобалните промени на изменението на климата, стр.33-49.

Slavov, N. and V. Alexandrov. 1997. "Influence of the Global Climatic Change on Agroclimatic Resources in Bulgaria," *Comptes Reudue de l'Academie Bulgare des Sciences,* 50(2):31-34.

Slavov, N. and V. Alexandrov. 1998a. "Spring Crops in Bulgaria Damaged by 1996 Summer Drought," *Drought Network News* 10(1):4-5.

Slavov, N. and V. Alexandrov. 1998b. Използване на математични модели на агроекосистеми за устойчиво производство в земеделието, Сп. Почвознание, агрохимия, екология. т.XXXII кн.6,стр.15-16.

Slavov, N. and E. Ivanova. 1998a. Влияние на глобалните промени на климата върху земеделието, Сп. Земеделие бр.6

Slavov, N. and E. Ivanova. 1998b. Адаптация на българското земеделие към глобалните промени на климата, Сп. Земеделие бр.11

Slavov, N. and E. Ivanova. 1999. Мерки за приспособяване на земеделието към вероятните промени на климата в нашата страна, Сп. Земеделие, кн.2

Tomov, N., 1989. "A Study of Stress in Maize," *Proceedings of the XXth Congress of EUCARPIA,* Götingen, Germany.

Tomov, N., N. Slavov and A. Gancheva. 1995. Засушаването и добива от царевицата през 1993 г., Сп. Растениевъдни науки, кн. 9-10, стр.47-52.

Tomov N., N. Angelov, N. Slavov, V. Alexandrov. 1996a. "Reaction of Maize Yield to Drought Stress in Bulgaria," *XVII-th Congress of EUCARPIA,* Thessaloniki, Greece.

Tomov N., N. Slavov and V. Alexandrov. 1996b. "Drought and Maize Production in Bulgaria," pp.169-176 in *Proceedings of the International Symposium on Drought and Plant Production,* Beograd, Yugoslavia.

Appendix 13.1

Stress response and grain yield of the hybrids from the 300-400 FAO group in 1993

Evaluation of the response to stress* (%). J = 20 July, A = 09 August; columns 1–6 where *1 = all leaves dry; 6 = all leaves healthy*.

Hybrid No	J1	J2	J3	J4	J5	J6	A1	A2	A3	A4	A5	A6	Harvested plants (number)	Ears (number)	Grain yield (kg/ha)
1			20	75	5				42	58			46	15	1451
2			17	83			19		38	43			42	18	1949
3				83	17				44	56			20	11	1159
4			16	80	4				21	62	17		23	15	1752
5			13	85	2			25	50	25			49	26	3010
6		66	27	7			2			87	11		52	22	2096
7			15	85					67	33			48	26	2427
8		9	89	2			24		25	49	2		31	19	1668
9				96	4				44	56			32	10	907
10		6	82	12	12				35	65			48	28	3110
11	2	8	83	7					50	33	17		47	29	3256
12		4	9	87						100			37	23	1740
13			20	80					49	44	7		46	26	2186
14		20	60	20					50	50			40	27	2750
15		10	85	5					65	22		13	49	39	3163
16		9	87	4					57	43			46	28	3297
17		6	83	11					31	31	38		37	15	1348
18		19	77	4				38		62			50	26	1950
19				85	11	4			42	53	4		40	16	1183
20		6	12	82						100			49	36	3185
21		21	64	15			21			64	15		52	16	1263
22		6	78	16			38		19	43			34	29	2393
23		10	87	3			10		40	50			51	20	2077
24			89	7	4				38	62			50	35	3335
25		9	85	68				34		52	14		53	26	2846
26		9	87	4					37	63			46	25	2233
27		4	76	15	5		56		24	20			52	33	2414
28		24	71	5					67	33			48	28	2541
29				80	20					100			47	16	1809
30	2	23	70	5			19		58	23			48	25	3109

* 1 = all leaves dry; 6 = all leaves healthy

Chapter 14

Drought Impacts on Agricultural Pests

Georgi Markov, Orlin Dekov, Maria Kocheva

As a typical steppe inhabitant, bound up with open plains, the common vole (*Microtus arvalis* Pall., 1778) is closely attached to agricultural areas. It has passed over from its primary natural habitats to agricultural ecosystems, where food sources are much more abundant. The common vole permanently inhabits Bulgarian agricultural ecosystems, forming colonies. It is the most numerous pest species among small murine rodents.

The economic importance of the vole increases simultaneously with the extension of areas under crop, because of wide fluctuations of their numbers and harvest wastage caused by them. During outbreaks the vole population reaches high density and causes enormous damage. In 1988 voles reached extremely high numbers, overrunning arable areas all over the country, causing incalculable damage. Such adaptive features as polygamy and female capability for feeding a large litter enable voles to breed intensively. The high reproductive potential of the vole (average litter size of 5-6) as well as their ability for monthly reproduction all year round under favorable conditions make them a species whose population numbers make them a subject of permanent consideration.

The common vole manifests marked changes in numbers every year, an example of species-environment interaction. Diversity in the changes in vole populations as well as factors causing them were and continue to be a field of investigation (Elton 1942; Frank 1957; Kratochvil 1959; Bashenina 1962; Straka 1967; Straka and Gerasimov 1971, 1977).

What causes such enormous fluctuations in population numbers does not yet have a clear answer. Nevertheless, common opinion is that dynamics of vole numbers depends on a complex of many factors, among which climatic conditions are considered to be of great importance (Hamar *et al.* 1964; Gladkina 1976).

Climate is a determining factor, but average parameters become more and more unreliable as area increases. At the same time, heterogeneity and asynchronous change in living conditions of the common vole over the whole territory of Bulgaria are determined by diversity of relief, soils, microclimatic features, vegetation cover and activities in particular agricultural ecosystems. Combinations of ecological conditions, which are differently favorable to vole reproduction and survival, arise over the same calendar year in particular parts of the territory. So localization is needed when population studies are carried out. The analysis of population dynamics should cover a region which is homogenous in regard both to climatic conditions and to structure and composition of the main agricultural ecosystems, inhabited by the vole.

The aims of our studies have been to reveal the character of the population development of the common vole throughout a relatively long period of time (1961-1991) compared with the vole life span, including a drought period (1982-1991); to carry out comparative analyses of the cyclic development of population dynamics characteristic under specific climatic conditions, including normal and drought periods in alfalfa and winter wheat areas voles inhabit; and to reveal climate impact on population dynamics.

Research Methods

The dynamics of the common vole population under specific climatic conditions in agricultural ecosystems in northeastern Bulgaria was based on development of the vole population throughout a period of 30 years (1961-1991) in alfalfa and winter wheat fields, localized in the Dobrich region. This is one of the regions where the common vole is most widely distributed and expresses high population density. There are 28,000-30,000ha of alfalfa and 130,000-150,000ha of winter wheat cultivated yearly in his region.

The state of the vole population in the Dobrich region was characterized for each calendar year by size of the alfalfa and winter wheat areas inhabited by voles and by their population density. This information was gathered for the entire cultivated territory of the region and recorded in annual records of the Regional Service for Plant Protection (Annual Reports 1961-1991). Our personal investigations (1982-1991) of vole population density, reflecting the spring and autumn period of population development, were carried out in specially chosen typical regions of regional agricultural ecosystems.

The initial results of these studies revealed the size of alfalfa and winter wheat fields where well-developed vole colonies were found. According to the number of colonies, vole population density was classified in the following groups: (a) single colonies; (b) 10-20 colonies/ha; (c) 20-50 colonies/ha; (d) 50-80 colonies/ha; and (d) over 80 colonies/ha.

The autumn observations carried out in September-October are reported in the present study. This is the time when voles complete their reproductive period and their population reaches maximum value under typical climatic conditions for the region.

In order to standardize the main population parameters of the common vole—density and size of the inhabited areas—and to transform them into comparable quantities in particular years and in different studied plots of the Dobrich region, the population parameter "coefficient of colonization (CC)" was used (Gladkina and Polyako 1973; Mihailova *et al.* 1982). This measure characterizes and integrates both parameters.

According to the specific way of sampling and the nature of data used (observation of all fields and recording of density in colonies per ha) in the present investigation, the numerical value of CC (average number of colonies per hectare of alfalfa and winter wheat fields in the particular year) was computed on modified equation:

$$CC = (20A + 50B + 100C)/D$$

where:

A is the size of area with low population density of voles – 20 colonies/ha;

B - the size of area with intermediate population density of voles – 20-80 colonies/ha;

C - the size of area with high population density of voles – over 80 colonies/ha; and

D - the total size of alfalfa and winter wheat areas.

The full cycle of vole population dynamics across many years covering sharply different values of its numbers was characterized by determination of numbers per calendar year in grades. According to common criteria for assessment of cyclic fluctuations in vole population development (Gladkina and Polyakov 1973; Gladkina 1976) as well as specific conditions of development and determination both of numbers and distribution of the common vole in Bulgaria, the following phases of vole population development were differentiated:

1. Grade 1 – (depression) – predominant low population numbers (from single colonies to 20 colonies/ha); only the main habitats (alfalfa fields) are inhabited;

2. Grade 2 – (numbers increase) – increase of density and enlargement of colonized areas in alfalfa fields, limited and low-density colonization of winter crops;

3. Grade 3 – (dispersal) – enhancement of colonization of alfalfa fields up to intermediate and high density and steady colonization of winter crops with low to intermediate density;

4. Grade 4 – (large-scale reproduction) – extensive colonization of alfalfa fields with intermediate and high density (over 50-80 colonies/ha) and increasing colonization of winter crops with intermediate and high density; and

5. Grade 5 – (reproduction peak) – maximum densities and sizes of colonized areas.

In order to find the relationship between the common vole population and climatic parameters during the study period, the microclimate in investigated ecosystems was characterized by its main components affecting development of biota (Tchernov and Penev 1993). The development of vole populations was analyzed based on the yearly and life span cycle, from October of the previous year to the current October. Throughout this period the climate in the Dobrich region was characterized (Agricultural Meteorological Bulletin 1961-1968; Agricultural Climatic Reference Book of Republic of Bulgaria 1968-1991) by average temperature throughout the period; precipitation sum; number of days with snow cover; and monthly sum of precipitation for the 12-month period.

The influence of climatic conditions on the population of the common vole was studied by stepwise linear regression with breakpoint-modification of the multiple regression analysis. The full cycle of development of common vole populations for normal and drought periods was compared using a standard descriptive method by duration, sequence of different phases of population development, average grade of the phase state and relative numerical level of population density in the particular phases. The analysis used the algorithms of the program WINSTAT (Statistics for Windows 1995).

Results

Population dynamics of the vole in Dobrich region during the period 1961-1991 is presented in Table 14.1. The variable course of population during the examined period emerges clearly (Figure 14.1). The difference between the minimum and the maximum values of vole numbers is 600%. The chosen model of multiple regression, which fits well to the biological conditions of vole development, reveals the high degree of interrelation between vole numbers and climatic factors in the region (Table 14.2).

The correlation between vole numbers and the main climatic parameters, generalized for the vole's annual life span (total precipitation sum, average annual temperature and number of days with snow cover) is high (R=0.94) and the deduced model complies with the real relationship between population and the examined climatic factors to a great extent (Figure 14.3). Detailed analysis of the interrelation between the population and the monthly precipitation (the climatic parameter that changes most sharply in drought) throughout the voles' life span also shows high dependence (R=0.98) and good correspondence between the deduced model with the real interrelation between these two parameters (Figure 14.3).

All results obtained clearly and unequivocally show that vole numbers in agricultural ecosystems are bound to climatic variation throughout their

Table 14.1 Population dynamics of the common vole during the period 1961-1990 in the Dobrich region of northeastern Bulgaria: 1–calendar year, 2–size of the alfalfa areas studied; 3–size of the winter wheat area studied; 4–state of the population, expressed by the Colonization Coefficient (CC); and 5–phase of population development

1	2	3	4	5	1	2	3	4	5
Year	ha	ha	C.C.	P.	Year	ha	ha	C.C.	P.
1961	19700	129900	0.00	1	1977	29500	133900	1.01	4
1962	20600	130200	0.00	1	1978	22000	135400	3.49	5
1963	11100	132400	0.10	1	1979	26300	130000	0.35	2
1964	21300	133800	0.26	2	1980	28800	130000	1.18	4
1965	19400	125900	0.37	2	1981	29600	131400	3.36	5
1966	19300	134100	0.49	2	1982	28200	134000	3.09	5
1967	20700	126500	1.49	5	1983	30800	149800	0.85	3
1968	19900	132900	0.51	2	1984	31000	149600	0.35	2
1969	23900	132400	0.23	2	1985	31100	149500	0.00	1
1970	26200	129300	2.66	5	1986	31500	150000	1.04	4
1971	27300	128900	3.40	5	1987	25000	142000	1.92	5
1972	26300	130300	0.72	3	1988	25400	149400	6.00	5
1973	29900	134200	0.35	2	1989	21200	147800	0.18	2
1974	34200	127500	0.85	3	1990	25400	148000	1.13	4
1975	33000	136500	2.10	5	1991	24500	147000	0.72	3
1976	31900	134200	0.00	1					

annual development cycle, so an alteration in population dynamics could be expected during drought. Comparative analysis of vole numbers during the drought period (1982-1991) and the longer normal period (1961-1981) shows visible increase both in the average (148%) and in the maximum numbers (171%) with concomitant growth of 49.02% of the population median during the drought period (Figure 14.4).

The common vole is a species with well-expressed cyclic fluctuations over many years and phase variation of population dynamics. The full cycle of long-term dynamics of population is characterized in different parts of its area by different periodicity and scope of large-scale breeding. The variation type is determined by the population biological features of the vole, realized under specific climatic conditions of their habitat.

The analysis of numbers and spatial variations in vole population development in the Dobrich region throughout a period of 30 years, expressed by corresponding phase grades (Figure 14.4) reveals differences between drought and normal periods. There is a clear trend in population development during normal period. The cycle begins from the state of number increase (grade 2), reaches the level of maximum densities and size of

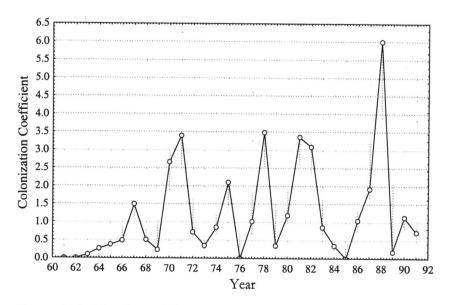

Figure 14.1 Numbers of the common vole, expressed by the Colonization Coefficient in the Dobrich region, northeastern Bulgaria, for the period from 1961 to 1991

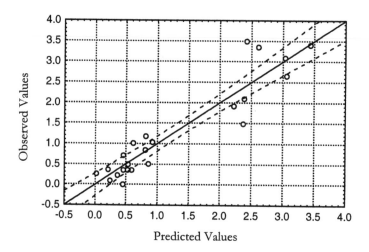

Figure 14.2 Observed versus predicted values of vole population and the main climatic characteristics in the Dobrich region (total precipitation sum, average yearly temperature and number of days with snow cover), generalized for vole yearly life span (with the 95% confidence limit of the regression; r = 0.94)

Table 14.2   Degree of interrelation between the vole population numbers and the main climatic factors in the Dobrich region of Northeast Bulgaria

Model: Complex linear regression Computation method: Quasi-Newton		
Dependent variable: State of the population numbers, (Y1 and Y2), expressed by the value of Colonization Coefficient (CC): R=0.98 Variation described: 87.75% Breakpoint: 1.257		
Independent variables:	Y1	Y2
R (precipitation)	-0.002104	0.008372
T (average temperature)	0.611449	-0.803757
S (days with snow cover)	0.012258	-0.009834
Const. BO	-5.94891	8.255006
Dependent variable: State of the population numbers, (Y1 and Y2), expressed by the value of Colonization Coefficient (CC): R=0.98017 Variation described: 96.74% Breakpoint: 1.286		
Independent variables:	Y1	Y2
AI (January ppt.)	0.005129	0.293488
AJ (February ppt.)	0.001172	0.325906
AK (March ppt.)	-0.005317	-0.092245
AL (April ppt.)	-0.002956	0.104283
AM (May ppt.)	0.002787	-0.085697
AN (June ppt.)	-0.003770	-0.252898
AO (July ppt.)	-0.001405	-0.118399
AP (August ppt.)	-0.001204	-0.028312
AQ (September ppt.)	0.002118	0.045088
AR (October ppt.)	0.005363	0.317335
AS (November ppt.)	-0.003195	-0.159234
AT (December ppt.)	-0.001853	-0.125867
Const. BO	0.735712	0.145478

colonized areas (grade 5), followed by decrease in numbers again to the phase of number increase (grade 2). The descending part of the cycle usually passes directly from the phase of reproduction peak (grade 5) to the phase of numbers increase (grade 2). The periodic sequence of large-scale reproduction is normally 3-4 years. The full cycle of vole' population development during drought takes 6 years. Its descending part, coinciding with the beginning of the dry period, is characterized by consecutive transition through all phases except phase 4 (large-scale reproduction). The level of maximum development of vole population density reached in the drought period is typically maintained for two years.

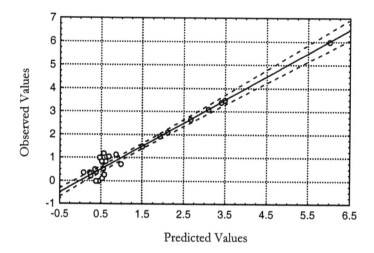

Figure 14.3 Observed versus predicted values of vole population and the the precipitation sum for each month of its life span in the agricultural ecosystems of the Dobrich region (with the 95% confidence limit of the regression; r = 0.98)

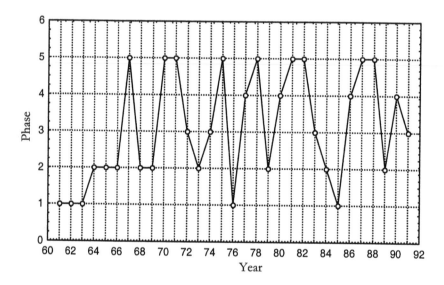

Figure 14.4 The common vole population state, expressed by the phase grade of its population dynamics in the Dobrich region of northeastern Bulgaria for the period 1961-1991

The type of full cycle population development during drought and normal periods differs not only by the cycle duration and phase, but also by the mean phase grade and by the relative numerical level of population density reached in particular phases (Figure 14.5).

Comparative analysis of variations of population density and phases of development of the common vole during the 30-year period as well as the character of population dynamics during normal and drought periods reveals a change in population development course which has occurred in the drought period. It is characterized by the clearly emerging trend toward raising and stable maintenance of a higher level of the population, enlargement of colonized areas and regions, restriction of the depression and prolongation of large-scale reproduction for two consecutive years. Such a reaction of population density development of the common vole during the drought period conforms to features fixed in evolutionary biological features of this steppe species, adapted to inhabit ecosystems with widely variable climatic conditions. The variation of climatic conditions in the Dobrich region is evidently within the adaptive reaction of species population density, and it has responded by intensive breeding and colonization of new territories.

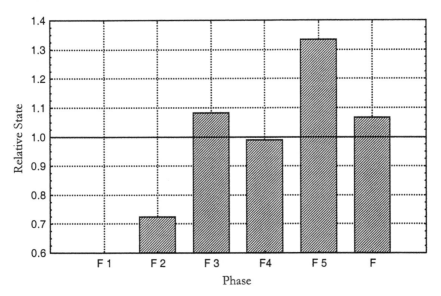

Figure 14.5 Relative state of the common vole population (numbers in the drought period/numbers in the normal period) in the particular phases of population development: F1 - depression; F2 - numbers increase; F3 - dispersal; F4 - large-scale reproduction; F5 - reproduction peak, F- the mean phase state in Dobrich region agricultural ecosystems during the period 1961-1991

Despite dependence of vole population dynamics on climatic factors, population cannot be considered as only climate-related. The influence of climatic factors is always refracted through the quantitative and qualitative state of the populations. Their population variations are a complex function of numerous endogenous and exogenous factors, and population vitality is determined by reproduction intensity, survival, interspecies relations, and resistance to diseases and poisons used in pest control. The cycles' periodicity and causes determining them could be different in different periods of time and in different places (Begon *et al.* 1980). Voles express quite labile, and moreover, cyclic variations in numbers, which are going on in a period of several years. Therefore, simultaneous study of the influence of biogenic and abiogenic factors on vole numbers is of exceptional significance for applied zoology. It could result in a scientific strategy for prediction and regulation of the numbers of these rodent pests both under normal and extreme climatic conditions causing considerable changes in ecological factors in agricultural ecosystems.

Conclusions

The course of development of the common vole in the Dobrich region established against a background of variable climatic conditions, including drought periods, represents an initial attempt to reveal the dependence of vole population dynamics on specific extreme climatic conditions in Bulgaria. The relatively high significance of a variety of climatic factors was found.

The considerably increased damage caused by harvest murine rodents and especially by the common vole in Bulgaria in recent years is undoubtedly also promoted by climatic changes in the whole country during drought. Probably as a result of changed climatic conditions, the percentage of successfully wintered voles increases, allowing the population to meet the spring with higher initial numbers. In combination with raised temperatures and earlier spring, this creates a more intensive and prolonged reproductive period and even active winter reproduction. Such winter reproduction was observed in 1987/88.

Examination of long-term dynamics of vole numbers in Bulgaria reveals a gradual growth of density and colonized areas as well as more frequent occurrence of vole outbreaks in recent years. In general, a trend to raising and steady maintenance of numbers at higher level, enlargement of occupied areas and affected regions, reduction of depressions and maintaining of large-scale multiplication for two consecutive years is outlined. From 1987 until now, almost every year the vole overruns at high density about 2 million decares under crop while short periods of depression, observed in 1989 and 1993, were quickly overcome.

An expansion of vole noxiousness is observed in regions where it had not caused considerable harvest damages in the past: Vidin, Lovech, Gabrovo regions. The development of a new region of strong pest infestation in northwestern Bulgaria has been confirmed. Thus large-scale reproduction has become more extended and frequent in the zone of potential noxiousness in North Bulgaria (Dekov 1997).

Recommendations

In order to respond adequately to the probable danger of vole outbreaks in agricultural ecosystems in case of a possible warm and dry spell in the country, taking into consideration the different degree of noxiousness and course of population dynamics in particular regions, the National Service of Plant Protection, Quarantine and Agricultural Chemistry must:

1. Organize stations for permanent specialized observation of vole populations in the Regional Services for Plant Protection, Quarantine and Agricultural Chemistry in Vratsa (Montana), Pleven, Russe (Razgrad), Dobrich, Sofia (Pernik), and Stara Zagora (Sliven);
2. Carry out usual minimum spring and autumn surveys in other Regional Services. During the period of higher danger, such surveys establishing the vole density have to be enhanced and more frequent;
3. Carry out vole pest control on the basis of an integrated management approach, mainly during the autumn phase of the population.

References

Agricultural Climatic Reference Book of the Republic of Bulgaria. 1971-1991.
Agricultural Meteorological Bulletin. 1962-1970. Hydro-meteorological Service, Sofia.
Annual Reports on Forecasting of the Regional (District) Services for Plant Protection, Quarantine and Agricultural Chemistry. 1961-1993.
Bashenina, N. 1962. Экология обыкновенной полевки. Moscow: Nauka.
Begon, M., D. Harper and K. Townsend. 1989. Экология. Особи, популяции и сообщества. v.2, Moscow: Mir.
Dekov, O. 1997. Обикновената полевка (Microtus arvalis, Pallas) в агроценозите в България - числена динамика, зони на вредност и средства за борба. Ph.D. Thesis. Institute for Plant Protection. Kostinbrod. pp. 38.
Elton, C. 1942. *Voles, Mice and Lemmings. Problems in Population Dynamics.* Oxford: Clarendon Press.
Frank, F. 1957. "The Causality of Microtine Cycles in Germany," *Journal of Wildlife Management* 21(2): 113-121.

Gladkina, T. 1976. Логическая модель динамики числености обыкновенной полевки в Калининградской области, *Proceedings of the Institute for Plant Protection*, 50:24-75 (Leningrad).

Gladkina, T. and I. Polyakov. 1973. Предпосылки многолетнего прогноза уровня численности вредных грызунов в Закавказье, *Proceedings of the Institute for Plant Protection*, 39 (Leningrad).

Hamar, M., M. Sutova and A.Tuta. 1964. "Certains traits foundamentaux de la dinamique des populations de rongeurs des agrosystems," *Bul .acad. sci. agri. et forest. SR*, 72(1):85-96.

Kratochvil, J. 1959. *Hrabos polni (Microtus arvalis)*. Praha: Nakladatelstivi Ceskoslovenska Akademie VED.

Mihailova P., F. Straka and I. Apostolov. 1982. *Plant Protection Forecasting and Warning*. Sofia: Zemizdat.

Statistics for Windows. 1995. Computer Program Manual, StatSoft. Inc.

Straka, F. 1967. Екология на обикновената полевка.. Sofia: Zemizdat.

Straka, F. and S. Gerasimov. 1971. "Correlations Between Some Climatic Factors and Abundance of *Microtus arvalis* in Bulgaria," *Ann. Zool. Fennici* 8:113-116.

Straka, F. and S. Gerasimov. 1977. Числена динамика и зони на вредност на обикновената полевка (Microtus arvalis Pall.) в България, *Ekologia* 3:79-91.

Tchernov, Y. and L. Penev. 1993. Биологическое разнообразие и климат, *Uspehi sovremennoi biologii*, 113,5:515-531.

PART VII
DROUGHT IMPACTS ON SOCIETY

Chapter 15

Bulgarian Society and the Water Crisis: Sociology and Ethics

Georgi Fotev

Both in scientific debates and everyday language, drought can be defined as a significant water shortage for different periods of duration. Of course, additional explanations of the phenomenon are necessary. The root of the Bulgarian word "susha" ("суша," *drought*) has two meanings. In ordinary language the word "susha" means earth, in contrast to water bodies (e.g., seas or rivers) or air. However, "susha" in Bulgarian also denotes climatic water shortage which may be the result of various causes but is usually due to lengthy periods with no rain. Because life without water is absolutely impossible, drought is seen as an intensive, fatal event, and in more extreme cases as a catastrophe.

Under the "Veil of Ignorance," or the Ocean of the Human Soul

Popular and precise quantitative scientific measurements of the importance of water for human survival are known, and it is worthwhile to recall them. For instance, water makes up 71 percent of the human body. Humans can live about one month without food but only one week without drinking water. Any human being needs on average 5 liters of water for drinking and cooking, and some 25 liters more for personal hygiene. It is well known that 97.5 percent of available water on the planet is saline, and the remaining 2.5 percent of water is to a great extent unfit for potable use. Therefore, only 0.007 percent of water reserves of the planet are both potable and easily accessible (Sadek 1999).

The living world of people in pre-modern Bulgarian society was very different and in some respects incomparable with the living world of modern Bulgarian society. However, it is indisputable that droughts have always interfered in people's lives, both in the past and present. The attitudes of people and society toward the phenomenon of drought and the significance they attributed to it are very different between the past and present. In a most general sense, water and drought are "experienced" differently in different periods of history. How traditional Bulgarian society experienced drought can be re-discovered in Bulgarian folklore, folk songs, myths, legends, proverbs, sayings, and tales. Droughts leave traces in people's lives and become part of their life experiences, which in turn become part of their artistic outlook on the world. Therefore, it is also important to consider the world of art, and to attempt dialogue between the diverse types of knowledge to reveal problems that otherwise would remain under the veil of ignorance.

As evidenced by the valuable notes of Dimitar Marinov who wrote from the end of the 19th and beginning of the 20th centuries, in Bulgarian folklore imagery, water ranks higher than earth. This was sanctioned by God himself, "for water is needed by everyone for everything." For Marinov, a man can live without bread for three to four and even more days, but without water he cannot endure even one day. Water is needed for land, forests, and cattle. Without water the earth cannot bear crops, and even if it bears something, it will not last but will dry up. Neither can cattle do without water. If water dries up, say old women who include Ivana from the village of Palamartsi, Popovo area, Roussa from Kipilovo, and Kalya from Kazuklise, "everything else will die" (Marinov 1981). These old women point out that water is not only used for drinking and irrigation but also for cleaning. "Without water, the whole world, the earth, the forest, the cattle, and man would stink." Water is the subject of devout reverence and ranks first in all rites and religious ceremonies. Water is viewed as sacred in most significant life events. When someone is born people turn to *prayer water*, after a funeral or commemorative visit to the grave *a libation is made*. At weddings a particular rite is related to *flower water* or "*tsvetna voda/ladvana voda*." Feasts are organized in honor of water. "Folklore forbids throwing garbage in the rivers or otherwise fouling waters and rivers" (Marinov 1981).

Magical powers were attributed to springs, rivers, and to rain. People believed that springs were peopled by fairies who appeared before humans, that fairies lived in rivers in the summer time, and that deep river pools were inhabited by devils and bad spirits. It was believed that men should not walk near these river pools at night, and if they had to cross a river, they should keep absolutely silent because otherwise danger was unavoidable. In a folk song, recorded in the Peroushtitsa region, Stoyan violated this belief and drew water from the fairy well. He saw there three fairies.

> One bathing a baby.
> Another cleaning her clothes.
> A third washing her head.

The fairies are described as naked women in long shirts. Stoyan disturbed and annoyed the three fairies. After returning to his home, the song relates further, he became feverish and died. In numerous legends such as this one, violations of humans were punished by death.

Water is treated not only through worship and reverence but also with hope and fear. Healing properties are attributed to water, as well as properties of moral purification. The extraordinary healer Petar Dimkov practiced and recorded water rituals of Bulgarian folk medicine (Dimkov 1977). He points out the curative properties of hot and cold water, of water procedures, and various types of bathing.

As mentioned before, folk beliefs also attribute moral purification to water. For example, if after a wedding it was found out that the bride was not a virgin, it was believed that she would bring disaster to the house. To prevent this she was taken to the river and even in the coldest weather was dipped naked and bathed, and thus the whole home was cleansed (Marinov 1981).

Sacrifice (*kourban* = offering) is an expression of the diabolic perception of water. Reverence, fear, and hope are combined in a mystical way. "Each river should take a sacrifice in the course of the year. For the Danube River there is a traditional belief that during spring flooding from melting snow, its level will not subside until either a sacrifice is given to the river, or the river will take one by itself, i.e. until either a live animal is thrown into it or a man drowns in it" (Marinov 1981). Sacrifice is a symbolic presentation of rules of the game played by human society (the community) and nature, where before taking you should first give.

In this game, the attitude toward rain is particularly interesting. The maker of the rules is unknown. The ways for penetrating folk mythology and religious feeling are indefinite. But there is no doubt that we are confronted with accumulated life experiences. Here is the record of Marinov (1981): "Rain comes from within the clouds and it pours rich crops in the fields and meadows and gives blessing to mankind. The rain is sent by God, therefore in times of drought they pray to God. At the same time, it is believed that the rain is given by St. Iliya; he is the one to decide whether there should be rain or drought. Dryness and drought come to reign only when God wishes to punish the human beings for their sins." A folk song says that "For three years no rain has fallen." Another says, "Nine years not a drop of rain has fallen." Marinov has also recorded a folk song about rain told by old Valkana from Antonovo, saying:

> Crosswise was the earth cracked,
> ...
> Burnt are the leaves in the forest,

Burnt is the grass in the field,
Dry are the rivers and springs,
Dry are the gorges in the mountain,
Great hunger on earth did come,
And then did the people repent,
Trust in the Lord they did and onto churches went,
Mourn did they and God beseech.
Pity did take St. Iliya himself,
The dark seas did he unlock,
Cloud and mist did he let down,
Quiet dew began to fall
And free-flowing drizzling rain,
Wetting the earth, the forest.

The rain is depicted as quiet and drizzling or loud and vehement. Rain around the day of the Savior is celebrated and welcomed with great delight as it foretells fertility and abundance of crops.

As Bulgarian society moved toward modernization, fundamental changes occurred in its world. Attitudes toward the changing world became secular and did not include magic, and the meaning of such elements of nature as water and its shortage was changing too. Consider the spring, which was earlier conceived of as "the most genuine symbol of life: forever coming and going..." The symbolic meaning of the spring is two-fold. It "combines in itself the power and diverse energy of the social and utilitarian principle. "Everything originates from water" is the inscription on a 1788 fountain in Balchik. The problem of building a fountain and the purpose of its own plastic design is closely related to mechanisms of tradition which have preserved artistic value and its existence in time..." (Lyubenova 1995). Because they were beyond magic and its secularization, the spring and fountain fell under a new horizon. This is the horizon of utilitarian thinking.

Modern society does not revere the natural processes of life but is instead motivated by governing them, while being completely unaware of the diabolic nature of such presumptions. Most sensitive to this dramatic attitudinal transformation are artists, who have insights about the integrity of the living world. Here, we think it is necessary to consider artistic perceptions of some of the most remarkable and talented Bulgarian writers over several generations: from Pencho Slaveikov to Yordan Radichkov. But let us first broaden the picture of Bulgarian folk magic through the eyes of its most enticing interpreter from the 1920s and 1930s, Nayden Sheitanov (1994a).

In the Bulgarian magical perception of the world, water is regarded as the pearled blood of earth. Sheitanov shows that springs were perceived as having social powers. A village's spring served as the place for ritual meetings of its young people. "There, the maidens and lads meet in the morning and evening like ancient nymphs and satires. The spring is the most divine of God's manifestations here down on earth. In a hidden gorge, somewhere in the mountain, the eagles bathe to wash away old age and become young again. The 'live' water heals the heavy wounds of the brave. The combination of water and sex in folk imagination resulted in 'a great synthesis of the existence'" (Sheitanov 1994b). Sheitanov speaks about "sexual cosmogony", "sexual geography" and "sexual geology" of Bulgarians, where the key focus is on water. "The sexual cosmogony in the beliefs of our nation," says Sheitanov, "is evidenced in songs and tales telling about the sun's wedding". The daylight sun travels in heavenly space as if only to look for a bride. Therefore, young women go to the spring for water early in the morning, "so that the sun would not see them." The sun is perceived as one big eye. The other saying runs with a characteristic wink: "The one-eyed (the penis) is our power." What is said about the sun is extended to include the sky which is "bending over, even lying on the earth, and from it fecundating seed falls down as rays, rain and thunder." The earth is likened to a woman: "woman without a man, the earth without rain cannot bear" (Sheitanov 1994b). The sexualizing of earth as a woman in folk sexual cosmogony is found in the sexualizing of geology and geography. This is obvious by the notion of the spring as vulva. Thermal springs are defined as the urination of earth. The anatomic metaphor can be inferred by names of springs such as "Chatal Chouchour" (Crotch Spout) (Sheitanov 1994).

When talking about "magic water, running from nine *chouchoura* (spouts), sexual metaphysics is again at hand. Sheitanov points out that the river bed is called a "womb" (Sheitanov 1994b). Such magical interpretations of conception reveal the multi-faceted significance of water for people's lives and intensifies its emotional value to people. Sheitanov's analyses reveal nuances of national attitudes toward what modern society now calls "water resources of the country." This culture was developed and enriched in national art and, as already mentioned, especially in national literature.

In his remarkable travel notes, the immortal Pencho Slaveikov (1972) reveals the deeply poetic nature of the Iskar River. Here is a fragment of what Slaveikov's pen has left for us: "The Iskar River is a grand traveler and still has not grown wiser in spite of the Turkish proverb that frequent journeys expand the mind. It takes its rise from Rila. He has his source in the hearth of Bulgaria and that is why he is so mutinous. This was whispered delicately to me by a playful wave that felt in a strange way that my heart is capable to understand the great truth."

"Old people relate that some hundred years ago Iskar sneaked upwards somewhere near Lyulin Mountain but then changed his route on a whim to take a short cut to pay a visit to the White Danube. He then courted the mountain Stara Planina and she took him in her arms, enchanted by the impetuous and passionate whispering of the cheerful handsome one" (Slaveikov 1972). If people of today could look at the Iskar River through the same eyes and with the same sensitivity, they would hardly allow the shifting of values toward aggressive utilitarianism that poisons the river with industrial wastes, dries it up and kills its "hearth". Utilitarianism's cold reasoning and rationality would reply contemptuously that everything we are talking about is merely poetic fantasy. But before answering this haughtiness, we might more intensely challenge it by quoting other poets that are well-known in contemporary Bulgarian literature.

The great poet Atanas Dalchev (1972) has dedicated more than one of his works to rain. In the poem "Rain," we read:

> Someone is noisily casting
> handfuls of seed-corn on the roof.
> Famished roosters scurry
> and start picking wildly

It is insane to try to turn poetic language and its metaphors into rational terms and schemes. The verses unveil invisible aspects of the experience of rain. In Dalchev's "Spring Rain" (1972) we read the verses:

> This spring rain falling,
> Only where a light is burning
>
> I forget the road, and the chalets;
> My sorrow do I forget
> And the cool coming from the heavens
> Washes away the worries from my face.

"Rila Streams" of Binyu Ivanov (1993) opens our eyes and souls to what we probably would neither see nor feel.

> Grass and sand blend in the sun
> Up the rocks, amidst the leaves you run.
> Your white voice shines from the crib,
> Waters, a river, the path to the sea,
> You babble and ripple; we live by your babble…
> You call from the dark, your caresses we hear,
> You babble and ripple; we live by your babble

The genuineness and depth of feeling revealed by poetry do not offer a resolution of social problems, contradictions and paradoxes we are facing, and we cannot expect to find such answers. Poetic vision can be regarded as a challenge and corrective against fragmentation of social life that exists in the current world of man.

In the 1980s, the prose of Dimitar Koroudjiev was opposite to the ideologically imposed attitudes and aesthetic tastes (or bad tastes, rather) of the time. This exceptionally talented writer drew attention of modern people to buried and forgotten sources of existence. Here is an excerpt from his remarkable novel (Koroudjiev 1998):

> Somebody behind your back turns the fountain in the country yard and with a cheerful sound the water gushes forth into the zinc bucket. But your inner rhythm of man, stars and pear tree remain the same. You evaded the race. And the sound of the jumping droplets agrees with it. An incredible feeling of security–nothing bad can happen to you until the water continues to flow. By this sound the approving life around you reminds you that it protects you.

We might say that this interpretation of the deep meaning of life and reality unveiled by the writer is beyond the limits of rational and instrumental reasoning.

In conclusion, let us look back to our exposition thus far. Let us examine the words of contemporary Bulgarian writer Yordan Radichkov on age-old traditional Bulgarian myths and legends that venerate water; myths which include elements of ballads and fear and that poetically address water–rivers, springs, and rain. In his original novel *Fear*, Radichkov offers brilliant narratives on the subject of water. In events which take place in a mill, a whole universe is uncovered before us. Radichkov (1980) says:

> I do not believe that even in the more distant past people believed in all kinds of told tales, although they told them at great length, in great detail ... gazing at the washer women above the mill, listening to the roaring water, rapt in the bubbling of the clappers, the people themselves became talkative, borne aloft on the wings of imagination, peopled the mill streams. The willow grows near them, marshes and swamps, the forests and gullies on the other side of the mill, the meadows and mountains with water spirits, vampires, wood-nymphs, fairy weddings, all manner of visible and invisible beings. They sheltered dragons in dry years in the sheep-folds and in their mulberry orchards and did all that not because they believed in it, but because they were keen on having all that, besides beans, lentils and maize, farm animals, silk worms, and a scraggy dog, and a bony woman ...

The writer is aware of people's need to go beyond material values and limitations; to go beyond what is empirical; to go beyond the world around them, which is within reach and accessible. Imagination played a role in people's world. In order to understand our world, we should look in the depths of the human soul. Yet, what is the human soul? And the writer says: "The human soul is an ocean, and you reader, how can you ever realize what is pulsing in the depth of that ocean?!" (Radichkov 1980).

There is a message in that story, and it is an important one: "The mill, instead of becoming a source of superstition, as it was in the old days, was most unexpectedly turned into a traditional roadside inn" (Radichkov 1980). We know what is behind this transformation. There are numerous similar "signs" in present-day Bulgaria. What has changed in current times is that many of the spiritual non-material values have been lost and things have been stripped of their many meanings. Tradition has become mere decoration.

Our conclusion should not be seen or interpreted as nostalgia or a romantic vision of the past. Such a return would be an absurd or comic hybridization of heterogeneous worlds. The bathroom or kitchen tap where chlorinated water often flows cannot be seen as a forest spring from which people drank; for a folk song says: thus, kneeling before the spring was not simply a physical posture–it was veneration before the spring. In short, we need a new ethic of water. We must liberate ourselves from one-dimensional utilitarianism. We need to critically evaluate our irresistible urge to dominate nature. We need a new societal attitude toward water resources. We need a new, far-sighted view of drought, which may have catastrophic consequences. However, in order to understand these needs, we should examine in greater detail the attitude of Bulgarian public toward the last drought.

The Totalitarian Project of Hyper-Industrialization of Bulgarian Society: Social Aspects of Drought

We already used the term *inclusion of Bulgaria* in the orbit of modernization. This was a process which cannot be outlined here in an analytical manner. However, we must note a fact of tremendous historical importance for Bulgarian society: a totalitarian communist system in Bulgaria dramatically forced the country into a grandiose and unprecedented experiment of the communist utopia. This utopia set as its aim hyper-industrialization as an alternative to modern hyper-modernization. The communist project in general aimed to overtake developed capitalist industrial countries, which presupposed limits to development, making it destined to fail. Moreover, the communist plan aimed to build a perfect society on earth and with it a new type of person. The foundational strategies of the plan were liquidation of private ownership, a basic transformation of social relationships, and unlimited domination over nature by doing away with society's dependence on nature. Moreover, it is noteworthy that the plan and practices it orchestrated have much in common with attitudes toward nature in the modern Western world. It should also be emphasized that the communist plan for Bulgaria was based on "science." The choice of the words "science" and "scientific approach" abounded in communist rhetoric.

The "hyper-industrialization" of Bulgaria, which prior to establishment of totalitarian socialism was an underdeveloped, agrarian country, was linked to major investments in science that were impressive for a country of its size. During the 1980s, a great number of research institutes and centers from the entire spectrum of science and technology began research and development efforts on a national scale to solve problems related to water and, more specifically, drought. There can be no comparison between reservoirs and irrigation systems built during the socialist period and the state of affairs prior to establishment of the Communist regime. A number of quantitative indices and parameters show that solutions to water problems in Bulgaria were far ahead of a number of countries during the 1980s. This is evident from other chapters in this book in addition to a great number of other publications.

The attention of the totalitarian government to water problems was systematic. Its approach to water issues created the feeling that society as a whole was actively involved in solving water problems. A complex plan for use and preservation of water resources in Bulgaria was drawn up in 1979 and then updated in 1986. Many researchers and experts from various scientific disciplines took part in compiling this document. The bulky text of approximately 500 pages is valuable in many ways, particularly because of its quantitative measurements, exhaustive data, and strictly empirical analysis. The motto of this work, as written by A. P. A Karpinski, a Soviet academician, is remarkable. "Water is the most valuable extracted resource. Water is not like an ordinary mineral resource in that it is not only a means for the development of industry and agriculture. Water is *a dual factor of the promotion of culture, it is the life blood* which creates life where it did not exist previously" (General Scheme 1986, italics added). This definition by itself can hardly be disputed. However, the context of the social system in which the interpretation was made is exceptionally important. The scientific character of quantitative measurements, data, and evaluations which the General Scheme contains should be truthfully placed within the context of the totalitarian system. And the true nature of this system is revealed by the remark on the cover, "For official use only." The General Scheme for Complex Use and Preservation of Water Resources was an official secret, and the public had no access to it. How then would it have been possible to consider water as a "dual factor for the promotion of culture," as the cited motto proclaims? The unbiased social scientist or analyst finds himself in a heterotrophic social world, where no conventional theoretical schemes, approaches, and cognitive instruments may function.

The totalitarian state and civil society are incompatible (Fotev 1992). It was only in the second half of the 1980s and on the eve of implosion and collapse of the system in Bulgaria that there was a very cautious public mention of civil society. It is outside the scope of this paper to consider what the sociological category "civil society" meant. However, we can illuminate it by citing examples. A totalitarian society does not allow the existence of any social group, any spontaneous social community, or any part of the population of a given region to express public disagreement and protest against decisions of the Party/State or any of their bodies regarding solutions to water problems, including planning sites for reservoirs, industrial enterprises, water pollution, allocation of water resources and determining any kind of consequence they may have. Free discussion of public plans and designs concerning development of the country was illegal. The behavior of public groups and society as a whole followed fixed rules and was under constant supervision and control wherever possible. The possible form of resistance was *passive resistance* of performers, who were aware that the instructions of totalitarian authorities were absurd, megalomaniac, and short-sighted. An anonymous team dominated everything and stunted one of the most important regulators of normal social relationships–individual or collective social agency–and led to irresponsibility when and wherever it was possible to avoid sanction. All this became a way of life. It stifled the fundamental motifs of socialization.

The so called "socialist ethics" and "communist morale" were the cause of deep decline of social morals and destruction for several decades of the moral basis of Bulgarian society. The totalitarian "hyper-industrialization" of Bulgaria caused negative consequences for the country's water resources and its measures for mediating problems commonly caused by drought. However, negative social effects were by far more dramatic and frequently absurd. Industrial production units built within the scale of the country which required large quantities of water were economically, socially, and technologically irrational and inadequate. Also, waste water from industrial production resulted in catastrophic pollution and exacerbated the negative consequences. Industrialization was also linked to a dynamic migration process, leading to concentration of the population and industrial units, quick growth of urban population, especially in larger cities. This concentration in turn created a number of water supply problems. In 1946, the urban population was 24%. In 1983, it reached 65%. The poor quality of the supply network increased the water losses. In fact, this was part of the typical irresponsibility produced by the system, which stifled personal motivation and any sense of responsibility. Moreover, as the system was not a market economy and water was considered to be "free of charge," criminal waste of water was common. Drought in the 1980s and first half of the 1990s made these problems more acute and magnified their subsequent negative consequences. A number of towns and regions were forced to comply with a strict water regime for many years, which disrupted daily life by not meeting sanitary requirements nor other domestic needs.

A variety of data could be provided on water balances and rainfall, demographic distributions, industry and agriculture, health statistics, etc. However, even the most carefully drafted selection and comparison of data has limited possibilities in helping us to understand the problems we are discussing. What we need are not simply statistical data, rather data stemming from relevant *sociological* empirical research in order to reveal empirical laws and relationships between various aspects of social life and the phenomenon of drought. Such specialized studies were not conducted under the previous regime. Sociologists were not included in joint projects, and in as much as economic aspects were examined, this was done within the framework of socialist political economy, which only took into consideration doctrinal party thinking. Some researchers and experts in the field of natural sciences were not blind to the evident social outcomes and conditions of their treatment of water problems. One of them recalls, "Thirty to forty years ago we were told that the culture of a people can be judged by the quantity of water consumed. The higher the consumption and the more water was used for sanitary purposes, the more advanced the development of industry" (Hristov 1995).

In the 1980s, the totalitarian system in Bulgaria, as well as the remaining parts of the Soviet bloc, found itself in deep crisis. A dissident movement grew on the eve of the collapse of the system. It is interesting and indicative that the first public clash after the ethnic crisis connected with the changing of Turkish names was between representatives of an informal environmental movement and the repressive state government in Sofia. The cause of the clash was the citizen's rejection of a government water resource project (Rila) and the government's manner of meeting water needs. Repression against a small group of courageous protesters resulted in the outburst of general discontent against the totalitarian communist system. This was the indication of the birth of a civil society which was to become a main agent in dealing with water problems of the country.

The Radical Transformation of Bulgarian Society and Formation of New Social Awareness and New Water Ethics

The implosion of the totalitarian system in late autumn of 1989 was an indication of the total crisis of society, which could only be overcome by radical democratic reforms and building of a market economy. What was needed was a transition to a modern democracy and modern capitalism. The transformation occurred in a hesitant and contradictory manner over a long period of time. By the end of 1996 and early 1997, when the former Communist Party which had transformed itself into a socialist party and was in power, once again found itself in the midst of a real catastrophe. The delay of privatization, including restitution of land, created conditions for a dangerous decapitalization of national wealth, which took various forms including crude plundering of property to money-laundering. The criminalization of society took on threatening dimensions. Irrigation systems and equipment, especially those used for agriculture, were either inoperable or badly maintained. The heavy financial crisis was a barrier to investments in hydro-construction work and for other solutions to water problems.

The peak of social tension and confrontation in connection with the drought was in 1994, continuing until the spring of 1995. The social conflict became civil disobedience and exploded when the Djerman-Skakavitsa project was started and a heavy water regime, a direct outcome of catastrophic draining of Iskar and Pancharevo reservoirs, was imposed on the capital Sofia.

The common correlates of these problems were changes in the demographics of society, including changes in mortality rate, birth rate, and natural growth rate. It is evident from Figure 15.1 that since 1990 the growth rate has been negative. There is no doubt that water shortages resulting from drought, as well as structural changes in water use, have direct, and in most cases considerable indirect negative impacts on life in general, and on the overall social atmosphere. All this was increasingly felt in regions where drought caused imposition of a heavy water-use regime.

The radical transition of society placed a number of key social problems on the agenda of water problems of the country and on the way water problems were treated, particularly when they became more acute during prolonged drought. A key problem was social justice in the distribution of water resources, which were seen as a public utility/public good. Once central planning and the command administrative system were done away with, the

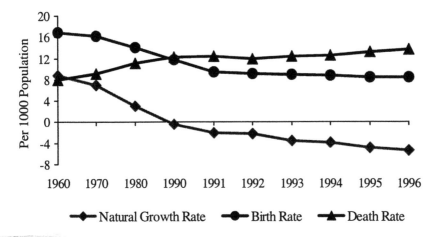

Figure 15.1 Dynamics of population in Bulgaria, 1960-1996

transition to a market economy began. The main question of the time was whether it was just to structure water use on a market basis like other spheres of life. In Bulgarian society there are supporters of this position. However, there have been increased discussions and recent confrontations between supporters of the inclusion of water resources as a market product on the one hand, and those who believe that the market is not a relevant mechanism for controlling water use and would lead to unacceptable social injustices. Support for the latter side of the argument has grown, just as it has grown in developed democratic countries. Here is one authoritative opinion: "Increasingly we are pressed to admit that water should become a commodity like any other market commodity. From this point of view the market could be used to regulate water supply needs to reduce its consumption and to resolve disputes between consumers within and between countries. This approach no doubt would increase the value of an increasingly valuable resource, and hence, decrease the waste of this resource. However, if market mechanisms function without any control, what would happen to the role of local authorities? How could they exercise their responsibility for implementing social justice for those who still have no access to drinking water?" (Major 1999). This most important dilemma before Bulgarian society calls for adoption of a new law. The regulatory potential of the market cannot be underestimated and neglected, neither should it be overestimated. The competence of the state and municipalities are well defined and a reasonable balance of interests should be established, giving ground for democratic discussions among democratically minded public. The problems related to the formation of a new culture and ethics where water is viewed as an exceptionally valuable commodity are by far more complex.

In pre-modern Bulgarian society we already explained the existence of traditional culture and ethics of water. One of the lessons that can be drawn from those historical experiences is the dual function of water: the consolidation of social integration on one hand, and emergence of disintegration and conflicts on the other. Yet the building of new culture and ethics of water in a contemporary society requires all members of the community to be well informed citizens. We cannot claim that Bulgarian citizens are well informed. A civil society has yet to be built in Bulgaria, and typical institutions of a civil society including non-governmental organizations have yet to emerge and strengthen. Liberalization as a key mechanism of democratic development and democratic decision-making has yet to be established.

When it comes to important decision making and construction of projects that have major social consequences, expert opinions hold significant weight in Bulgarian society. This was evident from the national sociological survey conducted by the firm MBMD during the scandalous construction of the Djerman-Skakavitsa project and the strict water regime in Sofia. When respondents were asked "Do you believe that when constructing major projects an environmental impact analysis on the consequences for the environment and public health should be mandatory?" 86% answered "yes".

In connection to social and political conflicts that had arisen, the survey showed that Bulgarian society as a whole and its people found it difficult to orient themselves and understand the arguments of parties in the conflict, while a considerable part of those with an established position could justify the arguments of both sides (Table 15.1). The majority of men supported the government's decision, while the majority of women favored a compromise solution, as they believed that arguments of both sides were justified. The position of women between the ages of 18 to 29 and 30 to 39 were similar, while the 40 to 49, 50 to 59 and over 60 groups were closer to the position of men. The percentage of people who were indecisive were similar for both genders and for all age groups. During times of severe crisis, it is very difficult for people to form rational opinions and to behave rationally.

The water crisis caused a stormy public debate in Bulgarian society and was focused on attitudes of social groups, political forces and the public at large regarding water resources and their utilization. Eminent scientists and water problem experts from the Bulgarian Academy of Sciences took a particularly active part in the debate. We must note that some experts defended conservative concepts and were working for interest groups at the

Table 15.1   Opinions about the derivation, Djerman-Skakavitsa (%)

Whose arguments are stronger?	Gender		Age (in years)					Total
	Men	Women	18-29	30-39	40-49	50-59	Over 60	
Government	32.4	24.7	21.0	21.6	32.2	34.5	34.5	28.4
Both sides	28.5	28.5	28.6	34.6	31.9	30.7	30.7	28.5
The inhabitants of Sapareva Banya	15.3	13.4	18.6	16.1	13.2	11.9	11.9	14.3
I cannot decide	23.9	33.4	31.8	27.7	22.6	22.8	22.8	28.8

expense of national interests. The leading scientists of the Academy, defending a dignified civil position, opposed the retrograde philosophy and group egoism of some circles of society, which were labeled by the general public as "mafia-like formations." At the same time, the dignified stand of academics drew public attention to the significance of water.

Water Problems and Social Planning

Social action in the complex Bulgarian society is subject to risk. It is not by chance that contemporary society is defined as a society of risk. Whether individual or collective, the agent of social action has limited knowledge. The more reliable the available knowledge and the greater the awareness of the absence of information, the smaller is expected risk and greater is rationality of choice. A reliable confidence threshold stands for predictability of the process. In comparison to natural processes that are governed by laws, social processes experience unexpected changes. Therefore, forecasts in social sciences and humanities are much more problematic than those in natural sciences. For this reason, it is traditionally accepted that representatives of natural sciences are skeptical of the scientific status of social sciences, which work with looser definitions and ambiguous terms. However, this is not the place to discuss irrelevant and incorrect comparisons of natural science and social knowledge. Cooperation between natural and social sciences on water resources, drought, and water problems allows for reliable social forecasts which aim to resolve social problems.

Chapter 18 documents changes in water use in Bulgaria. The changes are determined by the actions of various social agents in present-day Bulgarian society, by functioning of society as a whole, by power relationships between agents and political structures, and by economic and social relationships and so on. Water management must build a hierarchy of problems according to their social significance. From this point of view, Bulgarian society faces broad and intense debates on an international scale concerning social and moral problems of economic growth and the character of social development as a whole. This is the goal of sustainable development, which tries to prevent dangerous and future irreparable negative consequences for future generations.

Social planning should take into account the dynamic hierarchy of socially significant problems regarding water resources in the country; otherwise planning would be unrealistic and would lead to confusion of what is desirable with what is objectively possible. The definition of the hierarchy of socially significant problems of water use and social planning takes into account the democratic functioning of society and its main economic spheres, including effective functioning of a market economy. All this is extremely important considering the failure of communism which was utopian by nature. Social planning relies on a solid legal basis, clear legislation, sound research and predictions, and an appropriate water culture and ethic regarding water as the most significant public good.

As a rule, social planning in a complex society should be assigned to experts. It is necessary to clarify the task of experts with a view to specific social and moral responsibilities. Experts should outline potential positive and negative consequences of one step or another. They should outline the parameters of elections because it is a matter of policy to decide what course will be taken and how it will be accomplished. Political will and politicians' ethics of responsibility is exercised at this point. However, in a democratic society this means including the public in decision making; naturally, the public neither is nor could be fully homogenous. Various social groups can be differentiated in society with various interests, needs, goals, and concepts, including concepts of justice. Latent and open tension and conflicts in connection with the justification for use of a given commodity, or limitations on use of a given commodity, in this case water, are always possible. This is why the political elite have responsibility to reasonably manage tension and conflict in order to overcome conflicts. Good social planning identifies potential sources of social tension and conflict and assesses ways in which to overcome them.

Today and in the foreseeable future, social planning cannot be isolated from the process of globalization. The public is not informed about some international problems related to the Danube and major rivers leaving the territory of Bulgaria. This cannot be ignored in future. If no international coordination suitable to both sides is found, then there will be more complicated and conflicting issues in the future.

Conclusion and Recommendations

One cause of the crises of our times are the contradictions which arise between different types of rationality. In brief, scientific rationality determines social rationality. However, these two types of rationality do not work together automatically; what is economically rational is not automatically rational from a social point of view and vice versa. In order for social planning to be effective, it should be based on clear definitions. Yet they cannot be too rigid, so that they exclude or suppress spontaneity of social life. This is very important because social creativity and spontaneity are at work, in spite of the fact that social development is unthinkable without rationality, stability and a predictable order.

Social planning to ameliorate the water problem of Bulgarian society is a topic of the day, if we can use the term of the genius Max Weber. The more that scientists, politicians, and democratic societies are aware of this requirement, the more prepared they will be to meet challenges of the present and the foreseeable future of the country.

On the basis of this analysis it is pressing for Bulgaria to organize broad discussions about water resource use of the country with participation of experts from various fields and the legislature. National traditions should be reassessed and ways to develop a new culture and ethic of water should be sought. Water issues should be included in education, from earliest years to advanced education. Civil society should be transposed from having a passive position to an active one in decision making. Central and local governments should use social planning in the solution of water problems. To control excessive use and pollution of water, market mechanisms should be adapted to social goals and the potential of various social groups. The process of structural reform provides the possibility for technological innovation so that water can be used to its maximum economic potential with the least damage. Thus, water should be one of the major priorities in development plans for the country. Monitoring of social, cultural, and ethical aspects of water problems of Bulgarian society should be carried out systematically. With time, social sciences and the humanities will have an accumulation of empirical data that can be analyzed using scientific methods. Finally, the opening up of Bulgarian society in a regional, continental, and global context calls for international research cooperation.

References

Bell, D. 1973. *The Coming of Post-Industrial Society, A Venture in Social Forecasting.* New York: Basic Books.

Bulgaria. 1996. Социално-икономическо развитие. Национален статистически институт. Статистическо издателство и печатница, стр. 154.

Dalchev, A. 1972. Балкон. Български писател, София.

Dimkov, P. 1977. Българска народна медицина. Природолечение и природосъобразен живот. Т. 1, Изд. на БАН, София, 1977, с. 28 и сл., стр. 285-287.

Fotev, G. 1992. Гражданското общество. Изд. на БАН. София.

Generalna schema. 1986. Генерална схема за комплексно използване и опазване на водните ресурси. КНИПИТУГА. София, 1986, стр. 3.

Hristov, T. 1996. Водностопански съображения при използването на водите на Рила за задоволяване на нуждите от вода на селищата в подножието на Рила и София. - В: Вода за София. Библиотека Екогласност. София стр. 212.

Ivanov, B. 1993. Поезия. Изд. Отечество. София, стр. 9.

Jaspers, K. 1995. Малка школа за философско мислене. Изд. ГАЛ - ИКО, София.

Koroudjiev, D. 1998. Градината с косовете. Домът на Алма. ИК "Бард". София, стр. 226.

Lyubenova, I. 1995. От извора се пие вода на колене. Български фолклор, кн. 1-2, 1995, стр. 74.

Major, F. 1999. "On a New 'Ethics of Water,'" *UNESCO Courier*, March, p. 3.

Marinov, D. 1981. Избрани произведения. Изд. Наука и изкуство. София. 1981, стр. 77-79.

Radichkov, J. 1980. Спомени за коне. Изд. Христо Г. Данов. Пловдив, стр. 110-111.

Sadek, H. T. 1999. Търсенето нараства, предлагането спада, *UNESCO Courier,* March , p.8.

Sheitanov, N. 1994a. Българска магика. Село, В: Защо сме такива? В търсене на българската културна идентичност. Съст. 4., В. Еленков, Р. Димитров. Изд. Просвета. София, стр. 251.

Sheitanov, N. 1994b. Сексуална философия на българина, В: Защо сме такива? В търсене на българската културна идентичност. Съст. 4., В. Еленков, Р. Димитров. Изд. Просвета. София стр. 282.

Slaveikov, P. 1972. Искър, В: П. Славейков, А. Константинов, А. Страшимиров. Късове. Земиздат. София, стр. 41.

Chapter 16

Influence of Drought on the Bulgarian Economy

Nikolay Chkorev, Stefan Tsonev, Stoyan Totev

Is it at all possible to achieve sustainable development in a country that is not only lacking in quality water resources, but is also ill-prepared for negative consequences of drought? Bulgaria is a country which, compared to European countries, is poor in water resources. Therefore, from an economic viewpoint, expectations that use of water resources will be a growth factor are well grounded. Unfortunately, the Bulgarian economy appears to be non-adaptive and cannot transcend the so-called "transition to a market economy" without adequate economic strategies. Keeping in mind our effort to find a place in the European Union, which requires us to compare ourselves to Western European countries, and the fact that the "transition" cannot be put off indefinitely, from an economic point of view we need all possible resources (especially water) to move forward. Any attempt to reveal mechanisms behind drought's manifestation will at least allow policy makers to develop a realistic strategy for sustainable development during conditions of increasing water deficits.

Influence of Drought on Main Water-Consumers in Bulgaria

The biggest users of water in the country are agriculture, hydro-electrical power station (HEPS) electricity, and the chemical, food, and machine-building industries. Climate conditions influence these economic sectors in varying degrees—agriculture depends on climate conditions to a greater extent than do industry and transportation. In this sense, how much impact nature has on the economy is determined by the ratio between industry, services, and agriculture in the gross domestic product (GDP).

A study of the long-term influence of climatic factors on the economy cannot be separated from the background of the whole social-economic development of the country. Until the 1950s, Bulgaria had been an agricultural-industrial country. Agricultural production not only served to feed

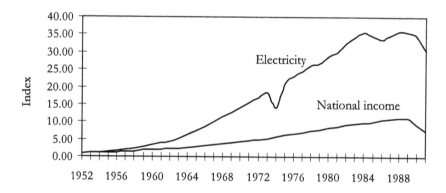

Figure 16.1 Indices of consumption of electrical energy and national income (1952 = 1.00)

Bulgaria's citizens, it was also an industrial resource. Since the 1950s, policies to accelerate the rate of industrialization have been implemented. For about forty years the economy functioned under non-market, centrally planned economic conditions. As in other East European countries, this period ended in economic collapse in the 1990s following political changes in 1989.

The mechanisms of planning were inherently in contradiction with all economic processes and phenomena that are variable, whether in the market or nature; the free market as a natural economic regulator and measure of comparative values practically ceased. For a considerably long period of time the use of value indexes for analytical purposes was difficult. During this period, policies were also implemented to accelerate industrialization. One of the objectives of this policy was to decrease the economy's dependence on nature by decreasing agriculture's share of the national income and by creation of an intensive and less nature-dependent irrigated agriculture. In order to ensure that industrialization was accelerated, the prices of agricultural production were kept artificially low and independent of actual supply and demand. For a general estimation of the processes occurring during this period, we used major indexes like total consumption of electrical energy and national income.[1] Figure 16.1 shows that for the period 1952-1991, the consumption of electrical energy in the country increased almost 40 times. The extent that this process has been economically beneficial can be seen from the fact that despite the increase of resource use, economic returns were sacrificed and national income increased by only a factor of 10.

---

[1] The cost of electrical energy shows the dynamics in the development of the industry; the national income is an estimation of the effectiveness of the economic activity in the country.

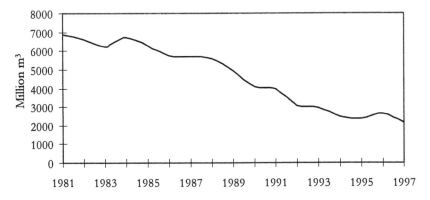

Figure 16.2   Fresh water used in Bulgaria (without hydroelectric or atomic energy production)

The influence of water deficits on the country's economy has not been documented with empirical data. The question is, to what extent does available information reveal the degree of influence natural factors have had on the economy? As a result of dramatic changes in the last years, including but not limited to privatization and increases in the price of power carriers, water consumption rates have changed. Figure 16.2 shows a clear decreasing trend in water consumption. In the period of 1981-1997 the water consumption in the country decreased by a factor of three! The decreasing trend during this time period is so significant that it can hardly be explained by natural water deficits alone. A considerable part of the water is used for production needs. Changes in both technologies as well as in the quantity of production lead to changes in water consumption rates. In the last decade the country has experienced a deep economic crisis that, because of decreased production, has caused a decrease in water consumption. But if the period after 1990 is labeled as critical, then how can we explain the fact that water consumption rates decreased at almost the same rate before 1990 as they did after 1990? The possible explanations are that water-saving technologies were introduced; the country suffered an economic crisis which affected water consumption; or the drought itself was contributing to decreased water use. The first hypothesis can hardly be accepted as significant; even if water-saving technologies were actively implemented, they could hardly decrease the water consumption at such rates. The more plausible causes for the sharp decline in water consumption are severe production declines and drought.

For the testing of the second and third hypotheses the influence of the confounding factors can be decreased if we study phenomena that are directly dependent on the presence or absence of water, for example, the volume of electrical energy production from HEPS. Even though reservoirs smooth over fluctuations in water flow, it can be expected that for a sufficiently long period of time the production of electrical energy in these power stations will depend not only on the installed capacity but also on the presence of water resources. Figure 16.3 contains data for the produced electrical energy from HEPS for the period 1939-1996 (*Statistical Yearbooks*, various dates). The production of electrical energy from HEPS increased until 1980, then decreased during the period 1981-1994, then increased again in 1995-1996.

If in the period 1939-1980 increases in production can be explained by the introduction of new power sources, the decline in 1980-1994 can hardly be explained by decreases in the installed capacity or by economic factors (for example, incapability to sell electricity). In this case, a possible reason for the decreased production of electrical energy from HEPS can be the lack of reservoir water resulting from drought. Second, if the water consumption in the country, the production of electrical energy from HEPS, and agricultural production have decreased not only because of economic reasons but also because of drought, then the data for crop production should contain findings similar to the aforementioned tendency. To check this hypothesis we used data on average yields in 1939-1996 (or the longest possible period after 1939) for the following water-intensive crops: maize, sugar beets, lucerne, soybeans, tomatoes, peppers, apples, melons, and watermelons. For each crop, an index is calculated based on 1939 or the first available year. From the individual indices for each year we calculated a general index of average crops for the year. The results are presented in Figure 16.4 which shows that the maximum values were reached in 1980-1982. After 1982, yields tend to decrease and then increase at the end of the period.

The population and the food and fodder industry are consumers in the economic chain extending from the farm. All changes in production should

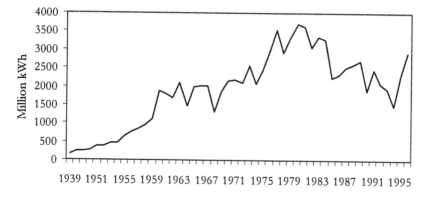

Figure 16.3 Hydroelectric production, 1939-1996 (million kWh)

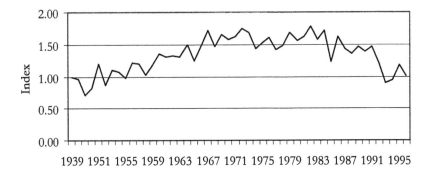

Figure 16.4   Annual index of the average yields (per decare) for a group of crops (1939 = 1.00)

reflect onto the consumers. Thus, based on data for the production of canned vegetables and fruits, wines, vegetable oils, oilcake, livestock feed, processed tobacco and mixed fodder, indexes were calculated based on 1939 data (or data from the first year in the data collection period). General average indexes are calculated from the individual indexes for each year. The results of these analyses are presented in Figure 16.5. The maximum crop yield index in both Figure 16.4 and 16.5 occurs in 1984. As a whole the crop yield indices increase until the 1980s, then decline until 1994, and then increase.

Total and per capita retail sales of fresh vegetables from 1952-1991 can be observed over two periods with a maximum in 1977. After that there is a decline until 1991 (Figure 16.6). More current data since 1992 has not been published. The change of the maximum in the data for the sales (1977) compared with the data for agriculture and production of electrical energy

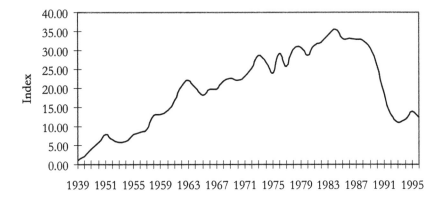

Figure 16.5   Average production index for a group of commodities from the food and fodder industry (1939 = 1.00)

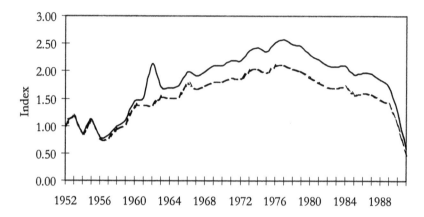

Figure 16.6 Index of fresh vegetable retail sales and fresh vegetable sales per capita (1952 = 1.00)

(1980-1984) is probably due to subjective-economic factors. Because of the absence of a free market and the surplus or the lack of agricultural production until 1982, retail sales of fresh vegetables began to decrease in 1978. In spite of this, sales of fresh vegetables cannot be independent from long-term agricultural production.

The index of economic activities studied here that are directly dependent on ample water resources (e.g., production of electrical energy from HEPS, production of agricultural products), as well as activities which depend indirectly on water resources (e.g., the production of canned vegetables and sales of fresh vegetables) show increases until the end of the 1970s through the beginning of the 1980s and then decline until the end of the 1990s.

This trend can be expressed with one-factor regression equations. Though there are more precise methods for presenting these dynamic processes, the sign before the regression coefficient in this one-factor model is the simplest way to describe the direction of observed tendencies.

Because 1980 is the approximate year when the direction of the observed tendencies change, regression coefficients were calculated separately for the period before 1980 and the period after 1981. Table 16.1 shows that the stochastic error at a probability of the regression coefficients of 0.95 is smaller than the coefficients, which allows us to assume that all regression coefficients are statistically significant. Table 16.1 also shows that the sign before the coefficients for the period before 1980 is positive and after 1981 negative, meaning that the tendency in all studied phenomena decreased before 1980 and increased after 1980.

Table 16.1 Regression coefficients showing the trends of development in the period before and after 1980

	Period till 1980		Period after 1981	
	Regression coefficient	Stochastic error	Regression coefficient	Stochastic error
Consumed water (without power plants)	-	-	-0.049	0.003
Hydroelectric energy	0.603	0.038	-0.528	0.162
Agriculture	0.023	0.003	-0.047	0.008
Food industry	0.846	0.037	-1.800	0.252
Vegetables sold per capita	0.048	0.003	-0.097	0.022
Rainfall 1939 = 1.00	-0.001590	0.001829	-0.01521	0.00536

How can the observed tendencies be explained? As mentioned earlier, the time period before 1980 had more water availability than the period after 1980, which was characterized by drought conditions. But increasing production in the period before 1980 can hardly be explained by favorable climate conditions. It is more likely that growth in production was a result of economic or technological factors (e.g., installation of new power plants, current development trends, etc.) and was not associated with climate conditions. On the other hand, the sharp turn after 1981 cannot be explained only by technological and economic factors because some power plants were put out of action and development stopped. There are probably other reasons which may include interference by unfavorable climatic conditions in this period. Figure 16.7 shows changes in rainfall 1945-1994. Rainfall amounts obviously decrease after 1980.

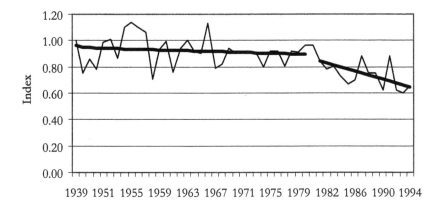

Figure 16.7   Index of rainfall 1939-1994 plus trends (1939 = 1.00)

This tendency is confirmed by the one-factor regression coefficients for periods before and after 1981. Table 16.1 shows that the regression coefficient for rainfall tendency is statistically insignificant before 1980, signaling that there is no clear trend in rainfall amounts that can be described with a linear model. On the contrary, the regression coefficient for the period after 1980 shows that the decrease in rainfall is statistically significant.

In this case, as far as the studied processes and phenomena are one-way dependent on such natural factors as rainfall, they will be negatively influenced. But here we have to judge whether in environmentally sensitive branches of the economy, such as agriculture, the drought will not lead to some critical threshold, followed by decline in many sub-branches. We fear that under current conditions of extreme shrinkage of irrigated areas the drought could not be ameliorated. In 1982 they were 11,851km² of irrigated land; today the experts estimate them to less than 3000km² considered usable for irrigation but in practice, even a smaller proportion are actually irrigated.

What we are most incapable of controlling is the deficit of financial resources. We cannot even fully use the average annual water flow of 2100 m³/capita, which is very small compared to countries like Norway, Switzerland, Italy and even Greece, because the built infrastructure fails to capture all of the water. As at a global scale, here too there is increased competition between the city and the village and as a rule, the bigger share of the scarce water resources is directed to the city.

Tracing changes in the macroeconomic indexes (like GDP) for the period of 1980-1996, we can clearly see that the agrarian sector is lagging (Table 16.2). At a base 1980 = 100%, the growth in GDP in the agricultural sector is negative for 1992. In dry 1993, the index drops to 72.6%. In 1996 the agrarian sector GDP did not reach 1980; the index is 75.3%. An elementary truth is that in countries with collapsing economies negative trends in GDP in one economic sector often negatively affect other sectors.

Bulgaria's agricultural share of its total GDP was 10.9% in 1990, approached Western European standards in 1995 when its value was 14.1%, and in 1997 it reached 26.2% (Table 16.3). These changes increase fears that possible future drought will lead to much more negative changes in the agricultural sector than in the shrinking industrial production and services sector. Even now, industry in general has not been excessively sensitive to drought, and, as was mentioned earlier, decreased water consumption makes it even more insensitive to drought. However, this does not mean that in the future influences of water deficit can be ignored—the development of the chemical and food industries are drastically affected by drought. The bread crisis that occurred a few years ago demonstrated how lack of flour can

Table 16.2 Growth of gross domestic product (GDP), 1980-1996

Sectors	Total		Agriculture and forestry		Industry		Services
Years	1	2	1	2	1	2	1
1980	100.0	100.0	100.0	100.0	100.0	100.0	100.0
1981	107.9	-	124.3	106.4	104.2	105.0	106.6
1982	112.5	-	134.0	112.0	121.1	110.0	88.1
1983	115.6	-	116.1	104.7	129.7	115.1	91.6
1984	122.8	-	134.7	112.0	136.9	127.3	93.6
1985	126.4	-	104.0	98.6	147.0	126.6	108.7
1986	133.5	-	119.6	109.9	158.9	131.9	96.8
1987	141.6	-	115.9	106.5	161.9	139.4	119.1
1988	148.7	-	118.2	107.8	168.6	143.7	128.9
1989	153.5	139.0	116.5	108.9	169.5	143.4	143.1
1990	139.5	-	112.2	104.9	148.3	125.5	136.9
1991	123.3	110.0	120.8	112.9	120.7	102.1	121.5
1992	116.2	105.0	111.5	104.2	112.2	94.9	117.9
1993	107.7	97.0	77.7	72.6	109.7	92.8	114.8
1994	109.6	99.0	85.3	80.3	116.6	95.6	111.6
1995	112.2	101.0	100.1	94.2	112.9	92.6	113.5
1996	100.1	90.0	80.1	75.4	101.4	83.0	103.7

1 – The indexes are in current prices until 1989, after that they are in constant 1989 prices;
2 – The indexes are in constant 1980 prices.
Sources: Statistical Yearbooks of Bulgaria for respective years and authors' calculations.

provoke huge difficulties in mill complexes, storage, and all aspects of pastry production. It is very difficult to say how much drought has contributed to problems in the chemical and food industry economies, but all industries that use considerable amounts of water will obviously be affected.

We also must be cognizant of possible negative social repercussions of drought that may be even more significant than agricultural impacts. Drought is a process that affects all humanity. If we know the intensity of drought and its sphere of influence on economic sectors, then we will be able to make responsible decisions that will minimize negative consequences. We are convinced that studying the multiple influences of drought and preparing for its negative consequences is an important task. Even in the worst case scenario, problem-solving tactics will ensure additional ground for maneuvering in the complex economic situation we are experiencing.

When we elaborate on the economic implications of drought it is necessary also to note the extreme difficulties in determining its contribution to the current bleak state of the country. Perhaps it is safe to say that drought's negative impact on the already dismal economy will be disastrous. What can we say about a country, which continues through a painful process

Table 16.3 Structure of GDP by sectors for some countries from southeastern Europe (in current prices)

Countries	Years	Agriculture and forestry	Industry	Services	Total
Albania[a]	1990	37.9	48.4	13.7	100.0
	1995	54.6	22.0	23.4	100.0
Bulgaria[b]	1989	10.9	59.4	29.7	100.0
	1995[j]	14.1	34.8	51.1	100.0
	1997[j]	26.2	29.4	44.4	100.0
Greece[c]	1990	16.0	26.5	57.5	100.0
	1995	14.3	23.7	62.0	100.0
Romania[d]	1990	21.8	45.9	32.3	100.0
	1995	21.7	39.3	39.0	100.0
Slovenia[e]	1990	5.2	41.8	53.0	100.0
	1995	5.0	39.0	56.0	100.0
New Yugoslavia[f]	1995	27.0	36.1	36.9	100.0
Macedonia[g]	1990	14.8	52.1	33.1	100.0
(FYROM)	1995	15.0	45.6	39.4	100.0
Croatia[h]	1991	9.4	31.4	59.2	100.0
	1995	10.3	22.7	67.0	100.0
EU[i]	1992	2.6	33.1	64.3	100.0

Sources:

a – Ministry of Finances of Albania (unpublished data)
b – Statistical Yearbook of Bulgaria, 1993 and Statistical Reference Book, 1996, 1998
c – National Accounts of Greece, NSSG, 1996
d – Statistical Yearbook of Romania, NCS, 1996
e – Statistical Office of Republic of Slovenia, 1996
f – EIU (1996-1997) Country Profiles
g – Social Product by Sectors. EIU (1996-1997) Country Profiles
h – In Constant Prices, from EIU (1996-1997) Country Profiles
i – Structure of the Gross Value Added in Constant Prices, EUROSTAT, 1995

of orienting national statistics to the real processes, rather than serving political interests? It was mentioned above that the drought is not subject to human control (or nearly so), thus dealing with it will encompass overcoming and compensating from drought damages where possible. Concerning this, we will trace phenomena and processes connected with those national economic activities and industries that, as a result of present and future circumstances, will continue to be directly and indirectly vulnerable to water deficit. We focus on agriculture and energy for the simple reason that currently they are two of the only branches of the economy capable of producing competitive goods. We can assume that Bulgarian vegetables, for example, will not be less shiny or nutritious than those in the EU, unlike the manufactures that both countries sell. The same goes for the other universal product, energy. In regards to energy, it is enough to state that Turkey and Bulgaria are still negotiating on prices, but if the price is fixed at US$0.035 per kWh, the problem with the influence of drought on energy will be magnified. Keeping in mind the stronger EU criticism concerning the safety of the nuclear power station (NPS) "Kozlodui" and the ecological requirements concerning the thermo-electric power stations (TEPS), it is possible that future substantial changes in the structure of the energy system will appear.

Bulgaria is a traditional producer of agricultural commodities. In the period 1935-1939 (1939 was the usual basis for comparison in national statistics for a long period of time) agricultural products and both food and tobacco made up 31.7% and 39.6% of the country's total exported goods, respectively. At the end of 1994, agricultural products' share of total exports is estimated at 18.6%, and its share of total imports is 9% (*Statistical Yearbooks*).

These substantial changes are due to a variety of factors. However, most economists agree that even though Bulgaria is now less of a rural country because of forced "industrialization," it still does not compare to even the average developed countries in terms of degree of industrialization. In other words, in the future the agrarian sector will play a significant role in our attempts to transform the declining economy. An analysis of the complex reasons responsible for the current state of the agricultural sector's decline is beyond the scope of this chapter. Rather, we will focus on the study of drought in this focal branch of the economy.

We will follow the changes in the average yields of two crop cultures that are influenced differently by drought, maize and wheat. Here it is important to know that the choice of these cultures is not accidental; maize is influenced considerably by water deficit, whereas wheat is comparatively tolerant. This is because the vegetation period of the maize can be divided into four sub-periods, each with different needs for water (Meranzova 1998) including the critical summer tasselling phase. Wheat is different because most significant vegetation growth occurs in the spring period which is characterized by water surplus. The changes in the average crop yields of maize and wheat in Bulgaria were traced for the period 1991-1996 (Table 16.4).

When we trace differences between average crops and water-intensive crops, the drought factor is decisive. Even in better general conditions, in dry years the average yields from all crops significantly decline. For example, in 1983 the average wheat crop yield was 320kg/decare and in 1981 it was 432.9kg/decare. For barley, a culture very critical to stockbreeding, there was also a decline from 370kg/decare in 1981 to 326kg/decare in the dry year of 1983. Processes for other cultures are similar. Quite curious is the fact that in

Table 16.4 Average yields of maize and wheat in Bulgaria, 1991-1996 (ton/ha)

Product/Year	1991	1992	1993	1994	1995	1996
Maize	5	2.8	1.9	2.8	3.8	2.2
Wheat	3.7	3.1	2.9	2.8	2.9	1.9

Source: Statistical Yearbooks 1992-1997.

these two years other factors of productivity, such as investment and fertilization, are more significant in the second year. Machinery and equipment exceeded the level reached in 1981; the aggregate power of the machines and working cattle in agriculture increased from 3920 thousand horse power in 1981 to 3944 thousand horse power in 1983.

The third most significant factor in plant growth after soil and water, fertilization, has not changed substantially. On the contrary, nitrogen was supplied more in 1983 than in 1981, some 550,285 tons (*Statistical Yearbook* 1984). But why does increased fertilizer application not compensate for drought? Simply, it is impossible for fertilization to replace moisture.

For water-intensive crops, the above rationale is even more important. Lucerne is quite indicative of this. First of all, unlike other crops, it has a long vegetation period and 2 to 4 harvests are possible in the first growing year. Specialized studies on the losses in this culture during an average dry year show that the damages are 17.2% at the first swath, 15.1% at the second, 10.9% at the third and 5.3% at the forth swath. In dry years the losses exceed 45%. It is clear that the average crop yields from lucerne hay vary substantially in dry years. For example, agricultural statistics show that average crop yields reached their lowest levels since 1960 in 1993 and 1994 (2800kg/ha and 2440kg/ha, respectively). It is enough to mention that it is possible to get 6000-10,000kg/ha in the first year and in the second-third year at 4-6 swaths to get 22,000kg/ha. Decreases in lucerne will in turn affect stockbreeding. Concerning damages from drought in the production of soy, sunflower and sugar beet, many years of observations undoubtedly prove that if we avoid the effects of drought, the additional crop yield for maize is 4250kg/ha; for soy, 970; sunflower, 620, for sugar beet, 16,020; and for fodder maize, 23,920 (Davidov and Givkov 1997). We should take into account the water deficit and ways for its mediation for opening up of markets in the East. For example, in 1993 for grain maize 5284 thousand ha sowed produced 1855kg/ha; the lost yield exceeded 900,000 tons.[2]

---

[2] The calculations are based on the differences in the average crops in 1993 and 1997. The first year is a typical dry one, and in the second year the average crop was 357.6kg/decare and the

If drought as a limiting factor in the development of fruits and vegetables can be reduced, it is quite possible that some social problems can be solved; these cultures require large labor input which could decrease the high level of unemployment. When we compare the data for cultures more tolerant to drought (in this case wheat) we see a different picture. There is clearly a tendency of significantly more gradual changes in the level of average crop yields. Therefore, if we accept the thesis that high crop yields are not necessarily the basis for highly developed agriculture but rather the achieving of long periods of stability, then the drought can be to a certain extent "avoided" with an increased share of drought-tolerant crops.

Influence of the Drought on Energy

In recent decades irrigation and the production of electrical energy have consistently been the main consumers of water in Bulgaria. In the period of 1981-1997 these two sectors have often changed their places as the leader in water consumption. Water consumption from HEPS reached a minimum in two of the driest years, 1985 and 1993. The diagrams show this clearly. It is natural for one to expect that the drought will increase problems in the production of electrical energy and this is confirmed not only by the existing data but also by expert opinion.

In 1989 HEPS had 17.7% of the installed capacity in the country but produced only 6.1% of the gross electrical energy. Experts unanimously agree that the role of hydroelectricity in the near and especially in the far future must increase. On the other hand, energy is a sufficiently competitive good on a global scale and the possibility of selling it to our neighbors is something that should not be ignored. Judging from data from the Central Control Office, for the period of 1980-1992 (from HEPS), the annual production of electrical energy decreased from 3713 million kWh to 2063 million kWh. Produced energy from HEPS as a percentage of total energy decreased from 10.6% in 1980 to 5.5% in 1995. The expectations concerning possible droughts are connected mostly with low load factors and decreases in the annual availability of the installed generation facilities. Experts state that the most unfavorable years 1985 and 1990 were a result of consecutive dry years in previous periods.

What are the economic advantages of the increased role of hydro-electricity? First of all, the price of water energy is partially independent of the

---

the total crop from significantly smaller sowed area–4.6 million decares–exceeded the total crop in 1993.

price of fuels. Water energy is replaceable and therefore preferable to other sources of fuel from an ecological standpoint. It also possesses some advantages in terms of energy-economic indexes. When taking into consideration the additional fact that the country has undeveloped hydroelectric potential that could feed about 87 HEPS and about 300 SHEPS (small hydro electrical power stations) with an average annual production of 6270 million kWh, it becomes clear that the influence of drought cannot be ignored. It is beneficial that this hydroelectricity potential is allocated over a considerably large area; if there are spatial variations of future drought, to a certain degree hydroelectricity will be less vulnerable to water deficit. Nevertheless, in all cases the drought has to be taken into account. In some cases, when production will be executed with low pressure and multistage cascades in the lower reaches of the rivers, the drought influences will be stronger; while in other cases, when HEPS and SHEPS are built in mountainous and semi-mountainous regions of the country, the influence of drought may be weaker.

Results from the operation of built HEPS have not been promising and show substantial diversions from their projected capacity. In 1980, the annual availability in 1980 was 1988 hours with an installed capacity of 1868 megawatts. In 1992 annual availability declined to 1068 hours with an installed capacity of 1975 megawatts. The drought also directly influences HEPS production because of the significant number of power stations working on run-of-the-river water (38 power stations).

Poor water resources of the country probably will force future hydro energy to conform to the needs of other water consumers, as they have in the past. But in a market environment it will be more and more difficult to satisfy the different needs of water consumers by imposing administrative decisions. In the new market economy nature will continue to provide water in the spring, irrigation needs in the summer months will again be mainly met by HEPS reserves, and in the fall and winter the hydroelectricity will not meet the needs for electrical energy. In a true market economy it is more likely that the difficulties that arose a few years ago, whereby water limits were imposed on the population in many regions of the country, will not occur again. Currently, not all HEPS work on a purely energetic regime; eight have mixed regimes (irrigation and energy), five have irrigation-only regimes, and five have water supply regimes. Even the cascades from "Batak Vodnosilov Pat", "Dospat-Vucha" and "Belmeken-Sestrimo" after their last stage satisfy substantial needs for water downstream. This classification shows that the consequences of drought will also be revealed in areas outside of energy. The effects of drought on the 70 HEPS with purely energetic regimes could be overshadowed by preserving and storing sufficient quantities of water from existing reservoirs. However, if unfavorable climatic factors are too strong or endure for long periods of time, negative consequences should be expected.

In purely economic terms, if we assume that 50% of the average annual production of hydroelectric energy (approximately 3130 million kWh) will not be produced after 2010 because of drought (with building of 87 HEPS and 301 SHEPS), we will lose millions of dollars from possible markets in Turkey. These potential losses can be even more substantial in severe droughts, and in the current state of the Bulgarian economy, as it was stated above, it is necessary to utilize all possible opportunities for filling the currency reserves of the country.

Influence of Drought on the Industrial Sector, Tourism, and Recreation

As a result of the total crisis in the Bulgarian economy in the last decade, many economic branches went through dramatic changes. "The crisis of the transition" left some industrial sub-branches without negative change. The 1997 GDP in real measurement (based on prices of the previous year) fell about 8% and annual income per capita was US$1200. The continuing negative rates of GDP change are naturally due to many complex reasons and it is naïve for one to expect that the influence of climate on this negative trend could be separated. As far as we can rely on forecasts concerning the development of industry, we expect that certain decreases in the gross value by branches of the economy will to a certain degree reduce the potential negative influence from drought. But this conclusion should be more specifically applied to chemical production and the cellulose-paper industry, whose indexes showed a positive development during the period 1992-1995. The status of the petrochemical and gas industries, whose indexes were 140 in 1995, is favorable. However, the food industry was characterized by negative rates; its index in 1993 was only 76.1. This is quite surprising, considering the fact that it was a dry year. In this case these results are due more to the poor crops than to the lack of water in the enterprises in this sector, but the initial reason is the drought, not other factors.

Production in many sub-branches, directly or indirectly dependent on the presence of water, reached absolute minimums in 1993. This year also registered one of the worst failures in the production of canned vegetables, only 86,779 tons (*Statistical Yearbook* 1998). But these results should be carefully interpreted. They may not be completely related to drought. We cannot assume that drought has played a significant role in the decline of production of water-dependent industries because data show even more significant declines in economic sectors with moderate and weak water dependence. The indexes of the industrial production of the food industry (46.7% at a base 1989 = 100%) are a considerably good achievement at the background of the general index of industrial production (48.5%) and even lower indexes of many other branches of the industry for the same year (*Statistical Yearbook* 1994). The drought factor in industry is significantly smaller than the other negative factors, but should not be ignored. In a more favorable general economic situation it is logical to expect that in strong water consuming industry, it is probable that drought would cause problems. The experience from the last years of the water crisis showed that although useful, efforts to find alternative water supplies (e.g., wells, underground waters) will not be able to fully compensate for the effects of continuing dry periods.

Concerning tourism and recreation, more concrete studies of the influence of unfavorable climate phenomena are needed. However, we can say that these are maybe the only spheres where we can expect different effects. For example, drought is a favorable event in the tourism industry for tourists in the Black Sea resorts wanting "nice" weather. There are regions in the world where hotels suffer losses from every rainy day. On the other hand, the absence of water in the bathrooms of hotel rooms has the potential to repel more foreign tourists than would a few rainy days at the sea coast. Drought and warming that effects winter sports could be important.

New approaches are also necessary to remedy the damage wrought by natural disasters provoked by drought. The human sacrifices and the huge material damages from recent fires in agricultural and forested areas directly and indirectly limit the economy and raise new problems.

Influence of Drought in Centralized, Transition and Market Economies

For different reasons in its modern history, Bulgaria has radically changed its socio-economic structure. Before 1944, Bulgaria followed the experience of capitalism in the Western world. This natural development with all its positive and negative aspects was drastically broken after World War II under an imported hegemony; command-and-control administrative approaches in the management of the country were imposed by force.

There is no doubt at all that in this time period many cardinal social-economic changes and the natural environment influenced the development of the country in different ways. The power of this manifestation to a great extent has been dependent on the structure of the Bulgarian economy, and, as is known, up to 1989, the country was a slave to the statement that the "right" way was the industrialization of the country. In this light, when we trace the possible influences of drought in this historic period, perhaps we should take into account the fact that despite its organic weaknesses, the planned economy possessed the ability to redirect resources from one branch to another and to some extent compensated for chronic imbalances.

Since 1989, the levers used in the recent past do not exist, because of the increasing private sector of the economy. Today dictating to producers what and how they should produce, with whom to trade and how to form their strategies for development is unthinkable. But in these new conditions, producers must deal with all the risks related to external factors. Drought is exactly such a factor for producers in the agrarian sector. Under conditions of resource deficits of any kind, compensation for consequences of drought is almost impossible. In the past such problems could be shadowed in the frames of the former socialist camp through adjustments in the provision of goods and services; now this is impossible. Lack of moisture cannot be compensated by increased irrigation to the so-called "irrigation areas" simply because they are no longer irrigated. Even though irrigation areas have been reduced, water is not available for more than 500 thousand decares. In other words the field for maneuvering in the conditions of transition economy is far smaller than a decade ago, and this means that the drought directly reduces productivity in the agrarian area.

To these factors we should add the fact that under conditions of the transition it is much more difficult to exert influence through the use of other factors (fertilization, soil cultivation, plant and biological protection) having in mind the continuing decline in agriculture and its decapitalization since 1989. But if this period ends and the schemes of the contemporary market economy are put into place, it is quite possible that with good management and planning, damages from the recent careless reactions to drought can at least be partly reduced. With good management skills, successful forecasting of drought can give new meaning to the necessity of pre-cultivation soil preparation, the choice and share of the water-intensive crops, the norms of fertilizing, and directions of their change. In order to assuage negative impacts of future drought during market economy conditions it is absolutely necessary to find ways to pay for supplies of water for irrigation, for increased costs of soil preparation—leveling and furrowing of the areas for gravitating irrigation, for example—for labor costs, for harvesting the additional production, and for increased costs of seeds and fertilizers. This approach can only be useful in a market economy. When we mentioned earlier about planning and the negative sense of this word, acquired in the conditions of directed economy, we should realize that in one sense or another this also exists in market conditions. Even if it includes activities for correct structuring of the sown areas under conditions of expected water deficit, to a great extent the losses will still prevail for the moisture-intensive crops. Proper planning can to a certain extent overshadow the lack of means, but each plan requires adequate finances. But such planning in new conditions must not be dictated, but proceed in a way that is seen as an instrument to bring about harmony to the economy. In time, planners should determine the links among, and influences on, the chain of global climate changes, whereby natural disasters could lead to negative economic changes which lead to social problems. History shows that planning for the future is becoming increasingly popular in contemporary market economies.

In Bulgaria it is inadmissible not to realize the comparative advantages that we possess in the agrarian area, even when there are droughts. In other branches of the economy, like hydroelectricity, the things will not be much different, and it will be more and more difficult to impose water use restrictions by force, whether water will be really used for designated purposes and HEPS will produce energy, or, like in past years, water for HEPS will be used for irrigation and drinking (or vice versa!). It is not difficult to predict that in a market economy it could become unthinkable that the installed capacity would not be utilized.

But how will a market economy actually utilize these inherited advantages? First, returning back to this type of economy will invalidate conditions in which "The State is a bad owner." The decision makers in a market environment must stop squandering resources and plan for potentially unfavorable effects from the deficit of natural resources, particularly water. This means that in order to at least survive they should be able to properly compensate for phenomena such as drought.

As history shows, only in conditions of a market economy will financial resources sufficiently be accumulated in order to import grain and thereby overcome water deficit problems for some period of time. Unlike the planned economy, where this is even slightly possible too, the long-term consequences resulting from the redirecting of resources will not be fatal. In contrast to planned and especially transitional economies, the market economy is able to effectively deal with such natural disasters without realizing declines. In any case, it is imperative that the economy of any country be in good general condition when unfavorable climate and natural disasters occur in order to avoid a total crisis. Therefore, we can draw the general conclusion that in conditions of a market economy effectiveness is highest. This is because in these conditions, energy and capital consumption per unit of production are at acceptable levels, and negative consequences from unforeseeable and uncontrollable events can be mediated.

Arguments about whether this can be achieved and to what extent the command economy would address such challenges are now purely academic. However, if we must compare the abilities of a certain type of economy to manage complications provoked by global climate changes, we can say without hesitation that the market economy is beyond comparison. Countries with a transition economy are challenged, but it concerns mostly their middle-term prospects, since sooner or later countries like Cuba and North Korea will have to join the market economy. Keeping in mind these reasons, with some certainty we can state that in the conditions of a command economy methods exist for reducing water deficit in some areas, but these methods compromise long-term socio-economic development. In other words, the market versus planned economy dilemma is imposed only in countries with exceptional ideology and generally the dichotomy is false. But in no case can we ignore the real problems of the transition to a market economy because a prolonged delay of this period may prevent us from achieving democracy. The Bulgarian society is tired and impatient; it expects the fruits of democracy today, not some unforeseeable time in the distant future.

But with these facts, how can we support the thesis that under the conditions of a market economy society will more successfully meet the challenges provoked by global climate changes? Figure 16.8 shows the changes in the levels of average yields of maize, a crop that was strongly influenced by the lack of sufficient water in Greece, Romania, and Bulgaria in the period 1991-1996. In Greece, where climate change influences are similar to those of Bulgaria and Romania, in the driest period (1992-1993) changes in average yields are insignificant. In other words, the economy of our southern neighbor has succeeded somehow in minimizing the failures in the agrarian

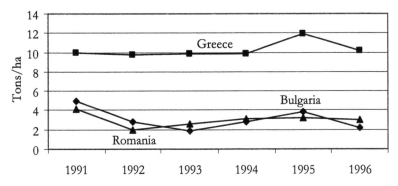

Figure 16.8 Average yields of maize in Greece, Bulgaria and Romania, 1991-1996 (tons/ha)

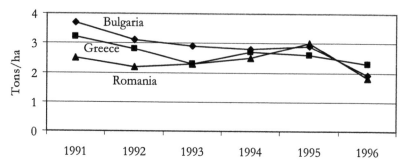

Figure 16.9 Average yields of wheat in Bulgaria, Greece and Romania in 1991-1996 (tons/ha)

area provoked by water deficit. But what is the picture in Bulgaria and Romania? The graphics shows that in the driest years the fluctuations in both countries are very large and in a negative direction. In Bulgaria production in 1993 compared to 1991 decreased drastically, the average crop yield of 5 tons/hectare decreased by more than 50%. In Romania, almost identical decreases are observed in 1992 and 1993. These declines in production characterize countries with undeveloped and declining agrarian sectors. Such sharp fluctuations illustrate the transitional economy's inability to adequately meet economic difficulties provoked by climate changes.

How in the same period did average crop yields from winter wheat compare? Here the differences among the three countries should not be so drastic. Climate failures should not be reflected in yield declines to the same extent as maize. Figure 16.9 shows that the tendencies in the three countries are identical. This confirms the conclusion that one possible way to reduce losses in grain-production in countries in transition to a market economy is to change the types of sowed crops. Since there are not enough mechanisms in place to compensate for water deficits, like irrigation, then crops tolerant to drought should have a bigger share in the agricultural sectors of countries like Romania and Bulgaria. This could reduce vulnerability, but in practicality the specific advantages these countries possess in normal wet years would not be used.

Previously, we studied processes concerning the adaptation to the market economy during the drought period. But how would have events developed in countries like Greece and Bulgaria under the latter's conditions of a planned (command) economy? Keeping in mind the aforementioned fact that there were ways to redirect resources administratively in order to overcome difficulties of the drought, we should expect that differences in the average crop yields of cultures sensitive to drought will remain, but perhaps will not be so extreme. Moreover, in the periods of 1969-1971 and 1974-1976, average wheat yields in both countries were similar. In the first period and second periods, Bulgaria produced 3910 and 3920kg/ha, respectively, while Greece produced 3070 and 3760kg/ha in these periods. When the drought began after 1980, we again observed considerable fluctuations in the average crop yields of maize in both countries. When we compare the situation in 1993 with the time of the planned economy, one can see that losses from the drought occurred, but they were not that extreme.

In spite of this, it is hardly wrong to say that these "successful" years during socialism prepared the country for the following economic declines. In the agrarian area one of the indicators for these unfavorable consequences was the extreme fluctuation in average crop yields of many crops in the last two decades. The extreme declines in crop yields for crops sensitive to water deficit in the last dry years in Bulgaria, when compared to the fewer declines in Greece, undoubtedly show that the problems arising from drought are extremely drastic in transitional economies. The only alternative for their mediation is a fast transformation of the country's economy to a real market economy. The quasi-market situation, observed in some years in the last decade, can only worsen problems, because in this situation the real interests of society are inevitably put after the interests of some groups.

General Conclusions

The economic aspects of the drought on different sectors of the socio-economic life of the country were ignored in the past but cannot be ignored in the conditions of a market economy. Analyses reveal that the negative consequences of drought will continue to be strongest in the agrarian sector (especially with regards to crop production), hydroelectricity and less so in some water consuming industries. There also exists many cascading indirect impacts in the economic system due to the role of water. Drought can also lead to weak positive effects in the recreation, tourism and agricultural industries, in the latter case through the inclusion of new, drained wetlands in cultivated areas. But this conclusion is valid for a very small territory of the country. These effects cannot compensate for the negative consequences in other sectors of the economy. The difficult situation in the economy of Bulgaria demands the use of all possible resources for development. In this sense the forecasting, mediating, and preventing, where possible, of the damaging consequences of drought are important economic tasks with large economic, social, and political effects. The market economy, unlike the "command economy," creates conditions that quickly reveal the impacts of the drought through market prices. It also creates conditions for faster and more timely reactions through the use of purely economic instruments and approaches for overcoming negative consequences.

Recommendations

Keeping in mind the importance of cyclic natural phenomena like drought for the country's economy, especially in the agricultural sector, it may be necessary to allocate funds designated for EU accession in this direction. A realistic approach requires plans for other natural disasters, like floods for example. The established state fund "Agriculture," which gives opportunities for opening credit lines, should support to a certain degree the restoration and preservation of irrigation areas that existed before 1989, thus partly compensating for drought damage. Even in transitional economies, good management, forecasting, developments, creation of proper infrastructure, and planning can lead to positive changes. That is why we recommend quicker adaptation to the EU methods of creating and applying agricultural policy, which implicitly addresses sustainable development and therefore counts deficits from natural resources. In the area of energy it is necessary to reevaluate the increasing role of hydroelectricity and initiatives concerning its exploitation. The "Strategy for the Development of the Energy Sector until 2010" should recommend measures to be taken during eventual dry periods. Efforts for avoiding the collision of interests emerging in the use of water for irrigation and electricity are also necessary. A new approach for forecasting and controlling disasters caused by drought is necessary. In order to solve existing as well as the future problems of drought, the state must allocate funds and resources for systematic observations and studies for understanding cyclical natural phenomena. This recommendation is especially valid for studies in the economic arena where the quantification is extremely difficult. A balanced approach in central versus and decentralized water use decisions is necessary to avoid the multiplication of mistakes.

References

Davidov, D. and G. Givkov. 1997. "Effectiveness from the Irrigation–Maize, Soy, Sunflower, Sugar Beet," *Agriculture Plus*, 1997, 4.

Meranzova, R. 1998. "Irrigation Regime of Maize," *Agriculture Plus*, 1998, 3.

Statistical Yearbooks of Bulgaria. Various years. Sofia: National Statistics Institute.

Worldlink. 1995. "Water–The Next Source of Trouble," *Worldlink*, November/December, 1995.

Chapter 17

Health and Hygienic Aspects
of Drought

Galina Gopina, Kosta Vasilev, Veska Kambourova, Ilian Dimitrov

Domestic Water Supply During Drought

The modern water supply infrastructure in Bulgaria was created in the 1970s and 1980s and over the last ten years has been gradually improved. Currently, over 98% of the Bulgarian population consumes drinking water from centralized water supply systems. These water supply systems have been approved by health authorities as having stable and acceptable water quality. Over the last few decades the pipeline water in Bulgaria has been used for drinking without any restrictions, in contrast to other countries. In compliance with sanitary regulations, water supply plants disinfect drinking water in order to insure the safety of consumers. According to the Bulgarian State Standard, disinfection treatment must guarantee a content of residual active chlorine in drinking water of 0.3-0.4mg/l after a thirty-minute contact interval between the water and the chlorine reagent (for details, see Chapter 9).

There are over 3000 water sources with permanent exploitation for drinking water supply in the country. According to the Ministry of Health, about forty-percent of the water used for drinking comes from surface water sources (e.g., dams and water catchment basins). The other sixty-percent originates from underground water basins (e.g., terrace wells, shallow and deep underground drilling, springs and drains capturing filtered rain water and shallow underground water). Although problems with water supply infrastructure have largely been solved, there are significant difficulties with the quantity of water available for drinking purposes, especially during summer months when large quantities are traditionally used for irrigation.

The main water basins for the drinking water supply in the biggest cities in the country consist of twelve dams (Figure 17.1), providing drinking water for 2.6 million residents, roughly thirty-percent of the country's 1993 population. In villages and towns, mountainous and semi-mountainous water basins, terrace wells, and shallow and deep drillings are mainly used, as well as large numbers of reservoirs that capture shallow surface and rain water.

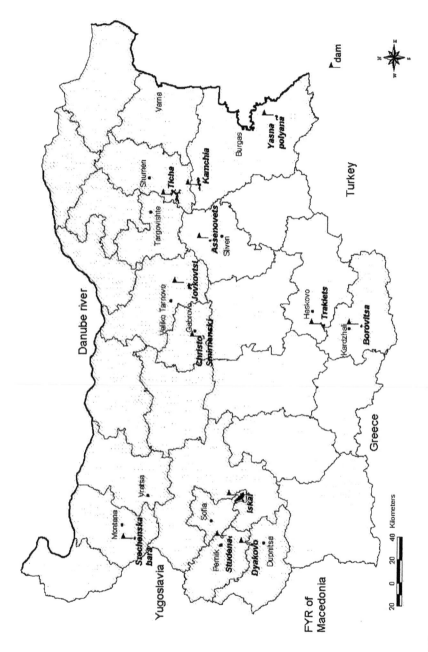

Figure 17.1 Major drinking water supply dams in Bulgaria

Karst springs, as well as drainage that captures rain water are the most vulnerable and rapidly affected by drought. Deep wells and terrace wells along the Danube and Maritsa rivers are the most stable sources of drinking water because of the sufficient amounts of captured water. Because of sufficient water quantities in the wells along the terrace of the Danube River, the towns and cities in North Bulgaria had no water shortages during the entire dry period. However, as a result of reduced water input for several consecutive years and untimely measures for economical consumption, a dramatic decrease in water storage volumes was observed in almost all dams during the dry period.

Storage quantity data presented in Table 17.1 were provided by the dam users. During this period, water storage quantities decreased sharply in eight out of twelve dams; some fell near their dead volumes. The minimal water amounts in the remaining four dams, which include Borovitsa, Assenovets, Yovkovtsi, and Ticha, significantly decreased in the driest years of the period, but there was no decrease in dead volumes. This is probably due to relatively

Table 17.1 Storage in major dams during the drought, 1983-1993
(million cubic meters)

Dam	Cities and towns	Popu-lation	Volume Total	Dead	Date	1983	1988	1990	1993
Iskar	Sofia	1113674	673.0	90	30.VI	407.3	553.9	328.6	315.7
					31.XII	332.4	403.2	190.7	147.9
Studena	Pernik	91075	25.2	2.4	30.VI	20.8	20.9	13.8	9.1
					31.XII	8.1	8	5.5	2.2
Dyakovo	Dupnitsa	41673	35	8	30.VI	18.5	17.7	19.6	20.6
					31.XII	12.7	7	10.8	7.1
Trakiets	Haskovo	81389	114	24	30.VI	47.7	41.6	26.4	29
					31.XII	32.2	28	26.4	20.6
Yasna Polyana	Burgas	198439	35.3	8	30.VI	12.4	15.7	12.7	15.8
					31.XII	12.9	17.3	7.3	12
Christo Smirnenski	Gabrovo	75999	18.7	4.2	30.VI	14.8	18.3	13.8	17.4
					31.XII	11.9	9.2	11.5	7.7
Ticha	Targovishte Shumen	139269	311.8	40	30.VI	174.7	223.8	116.8	271.2
					31.XII	169.3	160.6	95.7	235.1
Kamchia	Burgas Varna	505639	228.8	74.6	30.VI	151.7	164.3	52.1	192
					31.XII	125.2	108.3	44.3	144.2
Yovkovtsi	Veliko Turnovo	68059	92.2	9	30.VI	-	90.7	47.3	84.2
					31.XII	-	73.7	32.7	61.2
Assenovets	Sliven	106958	28.2	2	30.VI	-	19.6	12.6	21.9
					31.XII	-	14.5	13.8	14.2
Borovitsa	Kardjali	54216	27.3	4.6	30.VI	-	-	19.8	26.2
					31.XII	-	-	27.3	19.9
Srechenska Bara	Montana Vratsa	129613	15.5	1	30.VI	-	15.3	14.2	10.7
					31.XII	-	3.8	3.9	2.7
Total		*2606003*	*1605*	*267.8*					

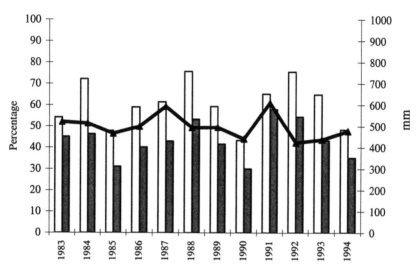

Figure 17.2 Dam Storage as a Percentage of Total Volume (first column, 30 June; second column, 31 December; solid line, annual precipitation in mm)

greater dam volumes as well as to water catchment areas located in mountainous regions with elevations over 1800m.

In only three years of the period 1983-1993 did 30 June storage volume in dams intended for drinking water reach 70% of total capacity (Figure 17.2). In 1985, 1990, and 1994 the storage amounts on 31 December decreased below 30%, following 2-3 consecutive dry years. The decreases clearly correlate with significantly reduced quantities of rain in the sub-periods 1983-1985, 1988-1990 and 1992-1993. The average quantity of rain per year during the dry period of 1983-1994 (499mm) was 15.2% lower than the annual rain during the period 1972-1982 (588mm). Only in two years during the dry period (1987 and 1991) did rain amounts reach the 1972-1982 average. In all the other years rainfall amounts were about 10 to 20% below average.

Water volumes in dams reaching or falling below the minimum quantity of stored water needed to maintain drinking water quality standards was observed in nine years. The situation is complicated by the absence of drinking water treatment plants for five of the dams where the water is only disinfected. The substantial decrease in quantities of dam water led to interruption of regular water supply. Many settlements were supplied only for a few hours daily, significantly deteriorating the hygienic conditions.

The water crisis was additionally complicated by the lack of adequate economic stimuli to control consumption of drinking water for other uses. As is normal in centralized economies, drinking water was supplied to the population at prices that did not adequately reflect its real costs. The effect of this mismanagement was that drinking water was used without restrictions by the public for irrigation in summer. The intensive summer consumption led to quicker depletion of the storage that was not replenished during dry periods. The situation was aggravated each subsequent year during winter minimums of storage quantities. Great water losses caused by pipeline damages further contributed to the depletion of water supplies.

As a consequence of prolonged drought, the greatest dams in the country, which had never before reached volumes approaching and below the dead volumes, were exhausted. Because of this, water regimes were imposed in many places where they were not normally applied, and they were extended in settlements where they were already implemented. Water regimes were especially implemented in large towns in the winter, when storage amounts reached critical declines. Thus, in 1989 and 1990, water supply regimes were imposed for two of the biggest cities in Bulgaria, Varna (supplied by the Kamchia Dam) and Burgas (supplied by the Kamchia and the Yasna Polyana Dams). Water regimes were strictly enforced in almost all other big towns that relied on dams as well, such as Pernik, Dupnitsa, Montana, Vratsa, and Haskovo, and eventually reached Sofia.

Analysis of Morbidity from Viral Hepatitis A during Drought

Hepatitis A virus (HAV) is one of the most stable viruses in the external environment. Its virulent properties remain intact in both water and solids for months. Epidemics usually arise from contaminated water or food, but the infection can also be spread by poor personal hygiene. The susceptibility of humans to the Hepatitis A virus is highest among children and adolescents. Once its victims have recovered from initial illness, they are immune to the disease for life. Since 1982 viral hepatitis A has on average accounted for about 75% of the total hepatitis morbidity in Bulgaria, similar to other countries. The infection is widely disseminated in cultures distinguished by poor socio-economic conditions such as unsanitary conditions, insufficient water supplies, and meager health care facilities (Iliev *et al.* 1994).

In the last 50 years the registration of viral hepatitis in Bulgaria underwent changes because of the development of basic knowledge in this field as well as the improvement of laboratory methods for diagnostics. Its registration under the name "viral hepatitis" began in 1951. In 1982 a separate registration of acute clinically manifested forms of viral hepatitis types A, B, and an undetermined type was introduced following the WHO classification (Gabev and Draganov 1986, 1988). In 1997 the monitoring of hepatitis was

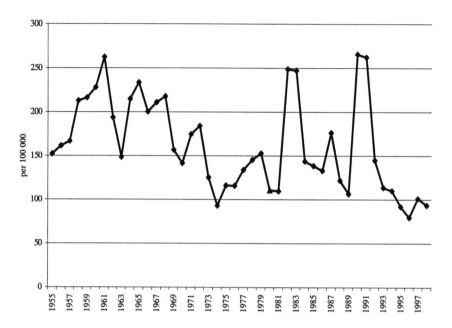

Figure 17.3   Hepatitis A morbidity in Bulgaria, 1955-1998

improved with the separate registration of hepatitis C and D, which had been distinguished from the undetermined type.

In the first years after beginning the official registration of viral hepatitis, the morbidity curve shows a strong positive slope, which is related to improved diagnostics and the new registration system. The peak morbidity rate of viral hepatitis occurred in 1961 (Figure 17.3). After this period, morbidity rate decreased. Epidemic elevations have been registered with intervals of 3-5 years (*Information Bulletin,* 1984). Since 1974, when one of the lowest levels of morbidity from viral hepatitis occurred, the inter-peak period increased to 7-8 years, from 1974 to 1982. In the second half of 1982 morbidity rose significantly and all the administrative districts of Bulgaria were affected. The incidence of disease from hepatitis A reached 249.3 per 100,000. High morbidity of viral hepatitis occurred in 1983 as well. Rates were especially high in village schools and nurseries (*Information Bulletin,* 1984). Between 1984 and 1990, excluding 1987, an inter-epidemic period with decreasing morbidity occurred. Gabev and Draganov (1988) claim that there were a series of epidemic outbreaks from viral hepatitis A in schools. The highest peaks in morbidity from viral hepatitis were achieved in 1990 and 1991. After this epidemic rise in the country, morbidity rates from virus hepatitis decreased.

Table 17.2 Areas with the highest VHA morbidity during the drought, 1983-1993 (per 100,000 individuals). Years of epidemic rise (1983, 1990, 1991) are highlighted (Y, yearly; P, periodic)

Area	Year					Years with morbidity > average	% population with water rationing
	1983	1988	1990	1991	1993		
Blagoevgrad	404	106	187	576	74	7	39.3-82.3 Y
Lovech	254	112	198	377	85	9	62.2-95.0 Y
Pazardzhik	179	194	160	282	153	7	16.5 Y
Stara Zagora	209	150	391	214	154	8	30.9-39.6 Y
Targovishte	558	78	365	610	44	6	50.7-56.5 Y
Kardjali	278	229	235	319	87	11	5.1-66.9 P
Sliven	229	169	334	298	269	11	7.3-68.9 P
Burgas	320	74	337	218	103	9	1.0-85.7 P
Varna	208	98	536	270	77	9	7.7-79.2 P
Shumen	268	113	700	317	82	6	5.3-92.2 P
BULGARIA	184	96	234	229	82		

Long-term studies in Bulgaria show that the morbidity rate of hepatitis A often rises between October and January and frequently peaks in November or December. Because of the seasonal autumn-winter prevalence of viral hepatitis A, morbidity elevation often occurred in two consecutive years. The two exceptional epidemic waves in the dry periods 1982-1983 and 1990-1991 were manifestations of the natural trend of the disease, strongly related to the level of collective immunity of the population. It is well known that the virus is disseminated intensively among children and young adults until the number of susceptible individuals is strongly reduced. The next epidemic occurs after a period of some years, when a new cohort of vulnerable people builds up and the level of collective immunity diminishes (WHO 1995; Thompson 1998).

According to epidemiological data, only VHA is influenced by water. Sample VHA morbidity data of the ten administrative districts with the highest rates during the 1983-1993 dry period are shown in Table 17.2. Data are based on infectious disease information recorded by the regional Hygiene and Epidemiology Inspectorates. We have previously documented VHA covering the country and provide only a summary here (Gopina *et al.* 2003).

The VHA morbidity data for the District of Blagoevgrad followed the national pattern during this period. In seven of the eleven years of the dry period, incidence of VHA in the District of Blagoevgrad is higher than that of the country. As seen in Table 17.3, the District of Blagoevgrad was one of the areas in the country with the greatest percentage of population subject to water supply regimes during the dry period (from 39.3% to 82.3%).

Table 17.3 Percentage of the population with regular water supply regimes during the drought (n.a. = no available data)

Area	Year				
	1983	1988	1990	1991	1993
Blagoevgrad	82.3	55.5	55.5	52.1	39.3
Burgas	69.2	26.2	65.5	56.7	1.0
Varna	23.3	78.9	79.2	11.2	7.7
Vidin	13.7	13.7	13.7	7.4	7.4
Vratsa	38.9	61.5	60.9	54.5	57.7
Gabrovo	n.a.	<5	<5	<5	<5
Dobrich	n.a.	n.a.	n.a.	n.a.	82.7
Kardjali	66.9	51.5	15.0	15.0	5.1
Kyustendil	n.a.	<10	<10	<10	<10
Lovech	74.6	63.9	80.3	62.2	83.8
Montana	n.a.	n.a.	n.a.	19.3	77.1
Pazardzhik	n.a.	n.a.	n.a.	n.a.	16.5
Pernik	n.a.	<10	40.0	40.0	<10
Plovdiv	<5	<5	<5	<5	<5
Russe	n.a.	<5	<5	<5	<5
Silistra	1.0	1.0	1.0	1.0	1.0
Sliven	68.9	68.9	7.3	7.3	7.3
Smolyan	n.a.	<5	<5	<5	<5
Sofia-city	0.0	0.0	0.0	0.0	0.0
Stara Zagora	39.6	35.3	35.3	35.3	30.9
Targovishte	56.5	50.7	50.7	50.7	50.7
Haskovo	5.1	2.6	2.6	2.6	2.6
Shumen	34.3	74.7	10.5	10.5	5.3
Yambol	33.6	19.8	19.8	19.8	19.8

In nine of the eleven years of the dry period, the VHA morbidity in the District of Lovech exceeded the country average. In 1987 an epidemic outbreak from contaminated water was observed in the town of Troyan (Gabev and Draganov 1988). The yearly water regime for the population of the area is one of the severest in the country, and encompasses from 62.2% to 95% of the population in various years. The VHA morbidity in the District of Pazardzhik was about average in the beginning of the period but in four years it was between 1.5 and 2 times higher than the average with a peak in 1992. In eight years the VHA morbidity in the District of Stara Zagora was above the average values. Between 30.9% and 39.6% of the population was subject to yearly water regimes. Morbidity rates in the District of Targovishte in the dry period shows very high values in the epidemic years 1983 and 1991. During the entire period about 50% or more of the population was subject to water regimes. In the District of Kardjali above average levels occurred during the entire dry period. Only after the implementation of a new water supply infrastructure in 1990 was there a significant decrease in percentage of the population subject to water regimes (5.1%-15.0%).

In the District of Sliven, VHA morbidity was also higher than the country average in all of the eleven years of the dry period. The number of people with water regimes was very high, especially in the first half of the period. In the District of Burgas during the dry period, morbidity was higher than the country average In nine of the eleven years during the dry period, the VHA morbidity in the District of Varna exceeded the average yearly values for the country. Between 18.1% to 23.3% of the population had water supply regimes, and during the years 1988-1990 when Kamchia Dam was depleted of water, this proportion reached 79.2% of the population. In the District of Shumen epidemic period of 1990-1991, rates were three times higher than the country. During six years of the dry period water regimes existed for a significant part of the population (18.8% to 92.2%).

Table 17.4 shows the areas with the lowest VHA morbidity in the dry period of 1989-1993. A tendency for lower morbidity in these areas occurred during both the inter-epidemic period and the epidemic years. As shown in Table 17.3, in five areas (Gabrovo, Plovdiv, Russe, Smolyan, and Sofia) a permanent water supply was available for about 95 to 100% of the population over the whole period. Only in the District of Vratsa was a water regime implemented for a significant part of the population.

Analysis of Shigelloses Morbidity during the Drought

Bacterial dysentery (shigellosis) is a disease which is caused by four strains of bacteria, namely: *Shigella shigae (Sh. dysenteriae), Shigella flexneri, Sh. sonnei* and *Sh. boydii*. According to the analyses of some authors (e.g., Gancheva *et al.* 1994; Gancheva *et al.* 1998), *Sh. flexneri* is the most prevalent strain in the rate of shigelloses in Bulgaria, followed by *S. sonnei*. The other two strains are only 1% of all strains isolated in Bulgaria. After the illness occurs in the victim, immunity is induced but is short-lived, making it possible for an individual to suffer from the disease several times. The dissemination of the disease is strongly supported by infectious carriers. In Bulgaria, this disease is mainly transmitted among children. For many years in Bulgaria, up to 70.6% of individuals who contracted the disease were children below the age of 14 years (Gancheva *et al.* 1995; Gancheva *et al.* 1996). Although the disease can occur any time of the year, it has a characteristic summer-autumn rise in incidence. The contagion arises predominantly by the ingestion of infected foods and

Drought in Bulgaria

Table 17.4   Areas with the lowest VHA morbidity during the drought, 1983-1993 (per 100,000 individuals; epidemic years are highlighted; Y = yearly; P = periodic; na = no data)

Area	Year					Years with morbidity > average	% population with water rationing
	1983	1988	**1990**	**1991**	1993		
Gabrovo	**n.a.**	99	**155**	**251**	53	3	<5 Y
Plovdiv	**149**	76	**197**	**144**	84	4	<5 Y
Pernik	**n.a.**	60	**68**	**73**	63	0	10.0-40.0 P
Vratsa	**273**	52	**171**	**284**	42	4	18.9-95.0 Y
Sofia-city	**82**	42	**146**	**80**	53	0	0 Y
Yambol	**137**	77	**131**	**188**	109	3	19.8-33.6 Y
Russe	**120**	81	**260**	**182**	28	2	<5 Y
Smolyan	**141**	62	**117**	**185**	93	1	<5 Y
BULGARIA	**184**	96	**234**	**229**	82		

water, and often arises in unsanitary conditions. Because of this, epidemic outbreaks quite frequently occur when water supply regimes are in effect.

The incidence of shigelloses over the last several decades has decreased (Figure 17.4). This decrease is undoubtedly due to improvements in the population's water supply as well as the increased prevalence of sanitary living conditions. Independent of the overall decrease in the total number of ill patients, the disease has diffused to different regions over the years. We have previously documented shigelloses nationwide and provide only a summary here (Gopina *et al.* 2003).

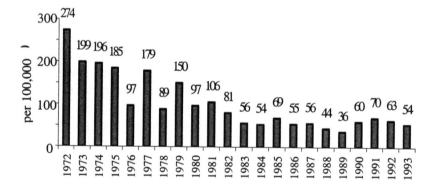

Figure 17.4   Morbidity from Shigelloses in Bulgaria, 1972-1993 (per 100,000 individuals)

Table 17.5   Areas with the highest Shigelloses morbidity, 1983-1993
(per 100,000 individuals; Y = yearly, P = periodic)

Area	Year					Years with morbidity > average	% population with water rationing
	1983	1988	1990	1991	1993		
Lovech	69	57	54	177	110	8	62.2-95.0 Y
Vratsa	76	98	126	203	105	8	18.9-95.0 Y
Dobrich	110	94	97	99	71	11	82.7-95.0 Y
Vidin	56	102	81	100	62	9	7.4-13.7 Y
Haskovo	70	39	73	76	57	10	2.6-5.1 Y
Silistra	61	78	92	97	93	7	1.0 Y
Kardjali	60	54	79	92	51	7	5.1-66.9 P
Shumen	56	40	66	112	96	6	5.3-74.7 P
Varna	87	66	50	58	99	8	7.7-79.2 P
BULGARIA	56	44	60	70	54		

The nine areas with the highest incidence of laboratory confirmed shigelloses during the dry period of 1983-1993 are listed in Table 17.5. The incidence of shigelloses in the District of Lovech exceeded the country average in eight years. Four years had especially high morbidity rates, and in two (1991 and 1992), the increase was more than twice the average. Over the whole period the greatest part of the population in the district (62.2-95%) was subject to yearly (mainly seasonal) water regimes. The District of Vratsa had a five-year period after 1987 with consistently high morbidity levels; it had almost yearly water regimes for large portions of its population. Over the whole dry period the incidence of shigelloses in the District of Dobrich was above the average for the country. It is well known that the District of Dobrich is one of the poorest in water resources. For this reason, water regimes are applied for most of the population of the district almost annually.

In nine years of the dry period the incidence of shigelloses in the District of Vidin was above the country average, yet a relatively small part of the population in the area (7.4-13.7%) was subject to yearly water regimes. In ten out of eleven years of the dry period the incidence of shigelloses in the District of Haskovo remained 20-30% above the country average The percentage of the population with yearly water regimes was relatively low (2.6-5.1%). In seven out of eleven years of the dry period the incidence of shigelloses in the District of Silistra was higher than the country average, significantly in six consecutive years, 1988-1993. High rates of shigelloses in this district have been linked to severe damages to the pipeline system and gaps in the process of disinfection.

In six years of the dry period, the incidence of shigelloses in the District of Kardjali exceeded the country average, and the highest morbidity rates occurred in 1987 and 1991. In most of the years more than half of the population had a water regime. In the first half of the dry period, the incidence of shigelloses in the District of Shumen was close to the country average with the exception of 1985. However, the morbidity steadily remained above average for the country during and after 1989 and was more than two times the average morbidity in 1989 and 1992. Water regimes were implemented in the area 1983-1985 and 1987-1989. The incidence of shigelloses in the District of Varna was barely above the country average in most years. Higher values in the beginning and middle of the period correlate with the low quality of the water supply during those years.

Discussion

Due to exhaustion of the great dams during 1983-1993, water supply regimes were imposed even in Varna and Burgas, reaching Sofia by 1994. Before the drought, water regimes had never before been implemented on this large a scale. Underground sources of water that supplied parts of the Districts of Lovech, Dobrich, Stara Zagora, Blagoevgrad, and Shumen were also affected by the drought. As a result, a high percentage of their population utilized water regimes.

Water supplies were most severely depleted during the peak years of the drought, 1984-1985, 1989-1990, and 1993-1994. As a consequence, water regimes were imposed upon a majority of the country's population. There is no doubt that deteriorated hygienic conditions caused by lack of tap water help the dissemination of gastro-intestinal infections, creating sporadic morbidity over the years. Water supply regimes interrupt conditions necessary for equal distribution of disinfection in the network. In these cases, agents of infectious diseases survive for a long time, creating conditions for both increased sporadic morbidity and periodic outbreaks of infectious diseases.

During inter-epidemic periods, VHA is mainly disseminated sporadically, but when proper circumstances occur, it is disseminated quickly in epidemic outbreaks. The current study has confirmed that rates of VHA and shigelloses have been consistently higher among large populations who live in administrative regions where there are insufficient amounts of drinking water. Each of the ten areas with increased VHA morbidity during the dry period had water regimes implemented for various durations of time for a significant part of the population. Basically all areas with adequate water supplies did not have increased rates of VHA morbidity.

In other regions with relatively good water supplies without water regimes, such as the districts Russe, Sofia, Smolyan, and Gabrovo, morbidity from VHA was below the country average. Even in the epidemic years the biggest cities of Bulgaria, Sofia and Plovdiv, where the total population is almost two million people and where a high prevalence of VHA morbidity could be expected, morbidity was in fact lower than the country average.

Since 1980 the morbidity of shigelloses in the country stabilized at a lower rate, and has even been decreasing recently. The morbidity rate in the dry period doubled in 1985 and 1990-1992. In six out of nine areas with the highest morbidity of shigelloses during the dry period, a significant part of the population was subject to water regimes. Most likely, factors other than decreased water amounts are responsible for the high shigelloses morbidity rates in other areas, including the degree of contagion, presence of susceptible child populations, awareness of the population, hygienic conditions of households and catering establishments, as well as frequent damage to water supply and sewerage systems. The investigation of these factors requires specialized epidemiological approach.

Conclusions and Recommendations

The prolonged dry period had an adverse effect on the capacity of water sources in Bulgaria. In 8 out of the 12 dams used for drinking water supply, water quantities reached critical values of dead volume during the dry period, and in some years storage volumes were exhausted to the sanitary minimum. Water regimes were enhanced and strengthened in regions with insufficient water resources or capacity of the water supply infrastructure. Because of the depletion of dams prolonged water regimes were imposed in some cities where previously there were no regimes. The presence of water regimes for the majority of the population in the ten areas with the highest prevalence of VHA and in 2/3 of the regions with the highest shigelloses morbidity confirms the significant contribution of insufficient quantities of purified and safe drinking water for the dissemination of these diseases.

To prevent negative health consequences in future long-term drought, it is necessary for local and central authorities to continue efforts to ensure proper surface and underground water sources with sufficient capacities on an annual basis. To prevent the depletion of drinking water sources, it is necessary to design a unique approach to provide adequate water yield from particular sources during different seasons. This approach needs to be based upon meteorological tendencies and upon the needs of the population. Local programs for rational consumption of drinking water must be elaborated, including implementation of restrictive regimes to avoid dams being depleted. Public health authorities must improve preventive actions such as disseminating health information, controlling the drinking water disinfection, regulating the treatment and preparation of foods, and promoting prevention strategies among vulnerable population groups.

References

Gabev, E. and P. Draganov. 1986. Анализ на разпространението на вирусния хепатит в Н. Р. България, *Служебен Бюлетин на НЦЗПБ, София* 2(5):4-20.

Gabev, E. and P. Draganov. 1988. Проучване върху възрастовото разпространение на вирусния хепатит по типове в НР България, Служебен бюлетин на НИЗПБ, София 4(1):24-30.

Gancheva, N. *et al.* 1994. Остри инфекциозни болести в България през 1993 г., *Информационен журнал на НЦЗПБ* 5:4-36.

Gancheva, N. *et al.* 1995. Остри инфекциозни болести в България през 1994 г., *Информационен журнал на НЦЗПБ* 4:4-32.

Gancheva, N. *et al.* 1996. Остри инфекциозни болести в България през 1995 г., *Информационен журнал на НЦЗПБ* 5:4-32.

Gancheva, N. *et al.* 1998. Остри инфекциозни болести в България през 1997 г., *Информационен журнал на НЦЗПБ* 4:4-32.

Gopina, G. *et al.* 2003. Здравно-хигиенни аспекти на засушаването, в И. Раев, Ч. Г. Найт, М. П. Станева, Засушаването в България. София: БАН, стр. 195-211.

Iliev, B., G. Mitov and M. Radev. 1994. Епидемиология на инфекциозните и неинфекциозните болести, София: Медицина и физкултура.

Information Bulletin. 1984. Информационен бюлетин на националната централа по вирусен хепатит в НРБ за 1983 година, МА, НИЗПБ, София.

Thompson, S. 1998. HepatoCite, International Hepatitis Update 14, Chester, Aids International.

WHO. 1995. "Public Health Control of Hepatitis A," *Bulletin of the World Health Organization* 73:1.

PART VIII
FROM IMPACTS TO CRISIS

Chapter 18

Water Resource Management During the Drought

Todor Hristov, Rossitsa Nikolova, Stiliana Yancheva, Nadejda Nikolova

Historical Review of Water Resource Management in Bulgaria

Bulgaria suffered considerable economic damage in the past because of frequent droughts and floods. For this reason, the first attempts to implement efficient water management strategies date back to the end of the 19th century, parallel to the formation of the new Bulgarian state. Several phases in the development of water sector activities and water management in Bulgaria are important.

Phase One includes the period from the end of the 19th century until 1944. It is characterized by an underdeveloped water sector and low water demands. State water policies were predominantly oriented to promoting water entrepreneurship and creating the structures and legal instruments needed for regulating water.

Initially, the right of "disposition on waters" was assigned to district governors (*Law on District Governors, 1882*). In the following years, acute water disputes erupted among rice producers and water mill owners. The disputes led to the development of the first economic regulator in this country, the "water right" (1887). Soon water management at the county level was evaluated as inefficient and was branded as incompetent and corrupt. For this reason, by virtue of the 1897 *Law on Public Works,* an "Inspectorate of the Hydrological Office" was set up with the Ministry of Public Buildings, Roads and Public Works. Its sphere of influence extended over water supplies, sewerage networks, and "water industry facilities." This marks the beginning of overlapping authority in water management. Certain authority was given to municipalities, so the Municipality of Sofia was able to grant concessions for the construction of water power plants (Boyana, Pancharevo) and for electrification. In the meantime, the 1891 *Law on Hot and Cold Mineral Springs* was passed with the objective of encouraging faster development of their business use, whereby their management was made the responsibility of the Ministry of Finance.

The *1904 Law on Property, Ownership Rights and the Servitude* dealt with a number of issues related to ownership rights and use of water. It outlined the rights and responsibilities of owners of plots that border water sites. For example, it listed the requirements for constructing weirs and dams and explained servitude rights. This law provided an effective legal framework for the field of hydrology and in essence is the analogue of the current laws in Austria, France and Great Britain.

The huge floods of 1897 and 1900 led to the creation in 1904 of a *Water Department* within the Ministry of Trade and Agriculture. In the beginning, its activities were oriented only to rice field irrigation and projects designed to protect against flooding along the Maritsa River, as well as to the reclamation of some marshlands near the Danube. Gradually, its activities expanded and part of it was relegated to the Directorate of Meteorology which at that time was under the Ministry of Education.

A document that played an important role in further developing water management was the report called *Water Economy, Its State in Some European Countries and How It Should be Organized in Our Country,* prepared in 1912 by Kisselkov (an engineer) addressed to the Minister of Public Buildings, Roads and Public Works. The report pointed out the advantages of water management of river basins and gave priority to water management that employed technology.

On the basis of the major recommendations of the Kisselkov report, in 1920 the National Assembly approved two important documents: *Law on Water Syndicates* and *General State Program on Water.* In actuality, both documents were aimed at promoting entrepreneurship in water affairs and specifically targeted collaboration of landowners for construction of hydraulic structures. Financial incentives, such as free grants, interest-free or low-interest credits, as well as mechanisms to make membership in water syndicates mandatory and to implement land parceling were envisaged for that purpose. *The General State Program on Water* was developed for 17 major river basins. It included a variety of measures, such as dam construction, river bed correction, stabilization and forestation of gullies, use of water energy, water use for irrigation and water supply, lake protection, drainage and reclamation of interior land marshes as well as along the Black Sea coast and the Danube River, and others.

The Law on Water Syndicates had a positive impact on water management by concentrating all water issues under the authority of one institution (the Ministry of Agriculture and State Property) and by creating effective regulation mechanisms (e.g., permits for water use, including restrictions under conditions of drought; "water right" charges according to the type of the water user, etc.). It also created the necessary structural units for surveying precipitation and runoff, elaborating water cadaster and registering water use points, issuing permits for and control of construction of water facilities, implementing water sector plans by river basins, and organizing, managing and controlling water syndicates. Water Sections were established as auxiliary units at the district level to assist the central management, and a Supreme Water Council was set up at the national level for management and co-ordination of all activities related to water management. Dozens of water syndicates were registered with the Ministry of Agriculture. They fell under two major groups according to their field of activity: those active in irrigation, drainage and river bed corrections and those active in the use of hydro energy.

After 1912 the Ministry of Public Buildings, Roads and Public Works also began to set up its district structures, Water Supply Offices that were later promoted to the status of Directorates. In 1935 the syndicates on water use for hydro-power were transferred (in compliance with the *Law on Electrification*) to the Ministry of Public Works. In 1938 an unsuccessful attempt was made to transfer the Water Department from the Ministry of Agriculture to the Ministry of Public Works as well.

A characteristic feature of the water policy of the state during that period was the creation of favorable conditions for development of the water sector through both preferential treatment and rights that were guaranteed by the legislation (*Law on Hot and Cold Mineral Springs, 1891; Law on Promotion of Local Industry, 1894; Law on Water Syndicates, 1920; Law on Land Irrigation and Drainage, 1940*). Specialized funds were also created for financing water sector activities (*Law on a Co-operative Water Supply Fund, Law on the Endorsement of a Fund for Land Irrigation and Drainage*). Despite the efforts made, however, the results achieved during that period were modest:

1. Water was provided to 1026 human settlements (about 25% of the total number of human settlements in Bulgaria) at an average water supply norm of 42 liters/capita/day;
2. Irrigation facilities were constructed for 37,500ha;
3. Approximately 25,000ha of marshlands were drained, 300km of levees were constructed and about 130km of river bed sections were corrected to prevent flooding;
4. 47 hydroelectric power plants (HPP) were constructed with total installed capacity of 47 MW and annual electricity output of about 168 GWh;

5.   Construction of the dams Beli Iskar, Kalin and Karagiol was started;
6.   Water consumption of about 0.4km³ accounted for about 2% of the total river runoff.

Phase Two consists of the period 1944-1987. This phase was characterized above all by large-scale hydro-engineering construction, centralized planning, budgetary financing and multiple reorganizations of water sector structures.

As early as November 1944, by virtue of a decision of the Council of Ministers, all water-related offices were merged into a Water Directorate within the Ministry of Agriculture. Special laws permitted the establishment of water syndicates for construction of the dams Stamboliiski, Koprinka, Topolnitsa, Beglika, and others. In 1949, the Great National Assembly approved a *Law on Water Economy*, whose principal objectives were to nationalize waters and their surrounding land areas; concentrate water management (with the exception of curative waters and those designated for water supplies) under the authority of the newly established Ministry of Electrification and Melioration where institutional units for design (Energohydroproekt) and for constructing hydraulic structures were created (Hydrostroy); and develop a "new state program on water" and ensure the implementation of a "comprehensive technical and economic control."

A Water Council was created with the Ministry of Electrification and Melioration for the purposes of resolving important issues related to water and its use. It was composed of representatives of the Ministry of Agriculture and Forests, the Ministry of Public Health, the Ministry of Industry and the Ministry of Public Services and Public Works.

By virtue of the *Law on Electric Power Industry* of 1948 all private power generation enterprises were nationalized, and in 1953 by virtue of a new *Law on Water Economy* all water syndicates were closed down and their property was nationalized. The law excluded from its domain waters used for water supply purposes, curative waters, transportation and fishing, as well as the waters of the Black Sea and the Danube River. The management of water for power generation and dam operations was left under the jurisdiction of the Ministry of Electrification, while the activities of state governance, construction of melioration facilities and economic management on water sector issues were transferred to the Ministry of Agriculture. The law canceled the charges for irrigation, drainage and protection against flooding and ruled that water use should comply with the regulations designed jointly by the Ministry of Agriculture, the Ministry of Electrification and the Ministry of Public Procurement and the State Reserve and should be approved by the Council of Ministers.

This was the beginning of a new redistribution of the institutional space of water management. Several follow-up laws and ordinances transferred functions of water management to other ministries (e.g., the ministries of construction, transport, public health, industry, chemical industry and metallurgy, defense), state institutions (e.g., The State Planning Commission, the Committee on Geology and Mineral Resources, the Bulgarian Academy of Sciences, the Civil Defense) and district, country and municipal councils. Each of these institutions developed its own structures for investigation, design, construction and operation. In 1960 the Institute of Water Economy and Construction in the Bulgarian Academy of Sciences, later renamed the Institute of Water Problems, was created.

Reorganization of the offices that collected water-related data was carried out as well. In 1950 a Unified Hydro-meteorological Office was set up with the Council of Ministers. It combined three former components: the Central Meteorological Institute, the Hydrographic Office of the Ministry of Agriculture and the methods of the Meteorological Office of the Army. In 1954 a Research Institute of Meteorology and Hydrology was established at the Hydro-meteorological Office. In the course of the same year both units were transferred to the Ministry of Agriculture, and in 1962 they were transferred again to the Bulgarian Academy of Sciences. In 1989 they merged and formed the National Institute of Meteorology and Hydrology. This institute was responsible for managing the state hydro-meteorological networks which was comprised of 163 meteorological stations, 373 precipitation metering stations, 33 agro-meteorological stations, 109 phenological stations, 236 hydrologic stations, 595 hydro-geological stations and one aerological station. Respective data were published until 1983. Since that time, access to them has become ever more difficult, and since 1991 they could only be obtained by purchase. The institutions and their subsidiaries dealing with water system operations have been developing their own hydrometric networks (labeled as "institutional"), the data of which are also hard to access and are not published. A large portion of them has by now been lost or destroyed.

By the end of the 1960s the majority of the large dams and hydroelectric cascades had already been constructed. Because the period 1948-1952 was evaluated as dry, a program for construction of small dams (a total of about 1,800 with a total capacity of about 0.6km^3) was implemented within a very short time period (1958-1963). However, this did not resolve the water deficit problem that was aggravated by water pollution resulting from the heavy industrialization and broad application of agricultural chemicals.

New legislation broadly began to develop new planning documents, programs and water administration. In 1963 the *Law on Prevention of Pollution of the Air, Water and Soils* was approved, and in 1969 a new *Water Act* was approved. In 1965 a supra-institutional body was established, the State Water Council, which reported directly to the Council of Ministers. It featured research and applied science units, a water inspectorate and a water sector cadaster. In 1971 the State Water Council was transferred to the Ministry of Forests and Protection of the Environment, and several years later, in 1976, to the newly created Committee for Protection of the National Environment. During that period an impressive number of by-laws were approved. By-laws were created for the following water issues: water quality, water use, water economy cadaster, the categorization of water courses, fines for water pollution, river-bed works, sanitary-protection zones around potable water sources, as well as more than 100 decisions, directives and other regulations aimed at resolving one problem or another that emerged in the meantime. Two types of planning documents were frequently developed, projects for managing water quality by categorizing various river stretches based on the needs of water users and the allowed concentrations of various polluting substances, as prescribed by specific ordinances and a Unified Water Sector Plan, whose objective was to meet growing water demands with new hydro-technical construction. Several types of measures were undertaken to combat drought, the first of which was used as early as 1982:

1. A new institution for the management of water quantities was set up, a National Water Council with the Ministry of Construction and Urban Organization (Decree No.13/1985 of the Council of Ministers);

2. A regulation was issued in 1985 that allowed construction of proprietary water sources. This resulted in the construction of more than 10,000 facilities without permits, mainly wells for use of groundwater;

3. During the period 1982-1988 five new dams with a total capacity of about $0.15km^3$ were constructed and commissioned into regular operation;

4. Some larger industrial enterprises were equipped with recycling water systems;

5. New planning documents were elaborated: a *General Scheme for Complex Use and Protection of Water Resources in Bulgaria* and a *Study on Maximum Catchment of Unregulated Waters with a View to their Utilization in the National Economy*. Their objective was to introduce maximum regulation of river runoff and water transferred from the transboundary rivers to the inland;

6. Construction of large hydro-engineering systems was started: the Water Supply Complex "Rila" for diverting the runoff of the Struma River to the basin of the Iskar River for Sofia City's water supply; and the Hydro-complex "Mesta" for diverting the runoff of the Mesta River to the "Belmeken-Sestrimo" Cascade, a hydroelectric facility in the basin of the Maritsa River (both the Struma and Mesta flow into Greece);

7. Preparatory work for construction of new dams in different regions of the country was started which included but was not limited to "Cherni Ossam" on the Ossam River, "Rayantsi" and "Iliina" in the Struma River basin, "Mladezh" on the Veleka River, "Byala" and "Neikovtsi" on the Yantra River, and four new dams on the Arda River.

These were the prevailing circumstances when *Phase Three* of the development of water management in Bulgaria started in 1988. It is characterized not so much by a change in the water policy of the state, but rather by its more active involvement with water management. It was during this time that the so-called *perestroika* was launched. Gorbachov came into power in the Soviet Union. He was the person who put a ban on projects that would transfer water from Siberian Rivers to the southern regions of the country.

A "Public Committee" was set up in Bulgaria to defend the city of Russe from transboundary air pollution. Water rationing was imposed on the water supply in the city of Blagoevgrad, because of the water diversion from the Struma River. The public revolted. Protests, blockades and civil unrest exploded in response to the Energoproekt's projects. The journalist I. Stamenov criticized those projects in his article published in the *Literary Front* newspaper in 1987. Later he submitted a report to the Central Committee of the Communist Party (who by then had ruled for 45 years), raising the issue of the infeasibility of the newly designed gigantic hydro-engineering plants (Stamenov 1988). The report was followed and supported by a statement from a group of researchers from the Bulgarian Academy of Sciences (Ignatov *et al.* 1988).

In November 1989 in Sofia, an international environmental forum was held, during which members of the semi-legal civil movement Ecoglasnost protested against the megalomaniac water projects (e.g., the Water Supply Complex "Rila"). The forum resulted in a conflict with the militia.

In 1990, nineteen representatives of academic institutes wrote a report on the environmental and economic infeasibility of the projects. In addition, the President of the Bulgarian Academy of Sciences (Sendov 1990) submitted a report to the National Assembly, on the grounds of which the construction of the Water Supply Complex "Rila" and the Hydro-complex "Mesta" were stopped by virtue of Decree No. 137/26 January 1990. The Parliament approved a new *Environmental Protection Act* banning the construction of new water diversion projects. In the meantime, as a result of the deepening economic crisis, water consumption diminished and the situation settled.

In 1991 democratic governments came into power and new people filled key positions in the state administration. Unfortunately, the majority of the new people in water administration had participated in designing the contested hydro-engineering complexes. Nevertheless, partial reorganization of the water sector began. The status of the National Water Council was changed–it was transferred to the Council of Ministers and its activity was subject to specific regulations. In addition, six regional units under the basin principle were set up. Work on drafting a new Water Act began. Thanks to financial assistance from the U.S. Agency for International Development (USAID), implementation of a pilot project for a Basin Council for the Yantra River basin began. However, because of strong opposition from state administration the project ended without success several years later (Mandova *et al.* 1998).

In 1992 Bulgaria was allocated a "water loan" in the amount of US$98 million. The World Bank developed a report describing the condition of the water supply system, putting emphasis on the large water losses and the need for comprehensive restructuring of water management in Bulgaria. The water loan remained unused for a long time because of the lack of adequate economic structures and laws put in place. In the end, a large portion of it ($US 49 million) was redirected to other sectors.

In 1994 a parliamentary and governmental crisis occurred and the governance of the country was assumed by a transitional government. At the same time, the capital Sofia City's principal water supply, the Iskar Dam, was in critical condition because of the drastic drop in its water level. As a result, severe water rationing was imposed on the capital. At that time the Government made the decision to proceed with the construction of the western water diversion Djerman-Skakavitsa, part of the cancelled "Rila" hydro complex for transferring water from the Struma River to the Iskar River basin. Despite the protests from the local population and 11 academic institutes, construction began (Hristov *et al.* 1995). The newly elected Parliament subsequently made the necessary amendments to the *Environment Protection Act* in order to legalize construction. 280 million leva were wasted because rain water destroyed this new derivation after being in operation for only three years.

The historical review clearly indicates that the frequent amendments in legislation and management did not contribute to the resolution of the problems, and Bulgaria has yet to find a successful model of water management. The future model of water management should take into account historical happenings, rather than repeating past failed solutions and errors. Public participation and public access to information are important prerequisites for the development of a successful model of water management.

Water Resources Supply and Use During the 1982-1994 Drought

In order to find an optimum strategy for water resource management during conditions of drought, it is important to know the unique conditions of water resources, water supplies, and water use systems during the period.

Water Resources

The potential water sources of the country include: the Danube River, the runoff of the inland rivers, and groundwater. Information about their distribution by type and consumption rate is provided in Tables 18.1 and 18.2.

The average annual runoff of the Danube River is estimated at 180km³. However, its use beyond the areas adjacent to the river banks is extensive. Its water is used for supplying long transportation facilities, which had huge water losses and required enormous energy for complex pumping systems. Existing facilities that rely on the Danube River use 4km³ of water for technological cooling at the Kozlodui Nuclear Power Plant and 0.8km³ of water for delivery of irrigation and water supplies.

The amount of fresh groundwater is estimated at 5-6km³, while the amount that is actually utilized (operational deposits) is estimated at 3.5-4.2km³. Since the groundwater is fed predominantly from rivers, precipitation, and surface runoff, it is strongly influenced by drought as well as by the rate and technologies of surface water use. The optimum quantity of usable groundwater is estimated at about 2km³ during drought. Therefore, existing facilities will most certainly use up all of this water resource during drought.

Runoff from inland rivers is characterized by considerable multi-annual, intra-annual (by months and seasons) and territorial variability. Inland rivers' average annual runoff amount is estimated at 19.5km³ and varies from 9km³ in a dry year to 32km³ in a humid year. More than 60% of the river runoff flows during the high water period (February-May), while only 3% of its

Table 18.1 Water supply and consumption (km³), 1983-1994 (Committee on Spatial and Urban Development, 1986, 1989; Knight *et al.* 1995)*

Year	1983	1989	1991	1992	1993	1994
1 Water supply less Nuclear Power (NP)	**10.087**	**10.626**	**8.726**	**7.023**	**6.012**	**5.808**
1.1 Surface water						
total	11.899	12. 859	10.589	8.718	8.936	8.798
river Danube	0.582	0.600	0.142	0.141	0.106	0.109
inland rivers, dams	7.103	7.934	6.512	5.605	4.243	4.221
others	0.780	0.580	0.560	0.302	0.576	0.463
1.2 Groundwater	1.622	1.512	1.512	1.274	1.305	1.016
2 Danube for NP	3.434	3.745	3.375	2.970	4.011	4.005
3 Water supply including NP	**13.521**	**14.371**	**12.101**	**9.993**	**10.023**	**9.814**
4 Water consumption less Hydro Power (HP)	**6.613**	**6.916**	**4.218**	**3.365**	**3.197**	**2.930**
4.1 Industry	2.090	2.047	1.512	1.367	1.592	1.438
4.2 Agriculture incl.	3.438	3.498	1.532	0.719	0.366	0.281
irrigation	3.438	2.768	1.204	0.571	0.280	0.239
4.3 Public water supply	1.502	1.659	1.490	1.507	1.379	1.310
households	0.357	0.408	0.404	0.374	0.366	0.358
industry and agriculture	0.583	0.530	0.470	0.317	0.224	0.172
other public consumers	0.296	0.316	0.157	0.236	0.192	0.171
water losses	0.266	0.405	0.459	0.580	0.597	0.609
4.4 Others	0.166	0.242	0.154	0.088	0.084	0.073
5 Water used by HP	3.484	3.709	4.475	3.601	2.816	2.873
6 Total consumption with HP	**10.097**	**10.625**	**8.693**	**6.966**	**6.013**	**5.803**
7 Water consumption by NP	3.434	3.745	3.375	2.970	4.011	4.005
8 Total water consumption with HP and NP	**13.531**	**14.370**	**12.068**	**9.936**	**10.024**	**9.808**
Difference (3)-(8)	-0.010	0.001	0.033	0.057	-0.001	0.006

*The data about water consumption by industry, agriculture, and the other economic sectors represent water supplied by the public water supply systems only (e.g., Water and Sewerage Companies). In order to avoid duplication of water quantities, water for industry and other applications was deducted from the public water supply in this analysis.

Table 18.2 Balance of water supply and demand (km³) during years of different moisture conditions

Year's moisture conditions at different probability	Humid 25%	Medium humid 50%	Medium dry 75%	Dry 95%
1 Water resources less Danube water for NP Kozlodui	**23.500**	**16.000**	**13.600**	**10.700**
1.1 Surface waters	21.500	14.000	11.600	8.700
Danube	0.800	0.800	0.800	0.800
inland rivers	19.200	11.700	8.880	5.400
re-used waters	1.500	1.500	1.500	1.500
multi-annual equalizers	-	-	0.500	1.000
1.2 Groundwater	2.000	2.000	2.000	2.000
1.3 From Danube for NP	4.000	4.000	4.000	4.000
2 Water resources total (incl. Danube for NP)	**27.500**	**20.000**	**17.600**	**14.700**
3 Water consumption less HP and NP	**4.821**	**4.621**	**4.421**	**4.721**
3.1 Public water supply	0.821	0.821	0.821	0.821
households	0.526	0.526	0.526	0.526
services sector	0.158	0.158	0.158	0.158
water losses 20%	0.137	0.137	0.137	0.137
3.2 Industry	2.100	2.100	2.100	2.100
3.3 Agriculture	0.900	1.200	1.500	1.800
including irrigation	0.600	0.900	1.200	1.500
3.4 Creation of multiannual reserves	1.000	0.500		
4 Hydro Power	**16.700**	**16.700**	**16.700**	**16.700**
5 Water consumption with HP	**21.521**	**21.321**	**21.121**	**21.421**
6 NP from Danube	**4.000**	**4.000**	**4.000**	**4.000**
7 Water consumption total (incl. HP and NP)	**25.521**	**25.321**	**25.121**	**25.421**
8 Difference (2) – (7)	+1.979	-5.321	-7.520	-10.712

Sources: Vodproekt, Vodokanalproekt, Energoproekt Committee for Spatial and Urban Development, State Planning Commission, Ministry of Agriculture and Forests, and National Water Council.

annual amount flows during the irrigation period (June-August) (Gerassimov *et al.* 1999). The runoff characteristics of the transboundary rivers Struma, Mesta, Maritsa, Tundzha and Arda are more favorable. Their watershed area covers 45% of the total area of the country and their water resources are estimated at 62% of the total runoff of inland rivers.

Taking into account the above description of water resource conditions during drought, the water strategy of Bulgaria is oriented above all to the regulation and redistribution of runoff from inland rivers. To this end a total of 2,035 dams have been constructed, featuring a total useful capacity of 7.2km^3 and gross storage capacity of about 8.5km^3. About 70% of the capacity of the dams is subject to seasonal storage and about 30% of it is intended for multi-annual regulation of the runoff, i.e. for its redistribution from humid to dry years. According to its designation, the regulated amount of water in the dams is distributed as follows: 21% for production of hydroelectricity, 21% for irrigation, 5% for potable water supplies, and 1% for industrial water supplies. The remaining 52% has complex multi-purpose designations; different combinations and ratios of the water are used for hydroelectricity generation, irrigation, and industrial and potable water supplies. Water use from dams is connected with numerous public conflicts because different consumers have different requirements with respect to the duration and quantities of their water consumption.

Numerous river catchments and more than 500km of water collection channels have been constructed for interbasin transfer and dam water recharge for hydroelectric generation, totaling 0.38km^3 of water. In addition, multiple water intake facilities are constructed in riverbeds, using recycled consumer water, including wastewater, and additional river water downstream from dams. The existing facilities allow intake, regulation and utilization of about 60% of the aggregate runoff of inland rivers.

Although the amount of runoff from inland rivers that is regulated and redistributed is relatively high (over 50% of total runoff), no comprehensive assessments have been made of the widespread negative impacts of hydro-engineering construction on the environment, the ecosystems, groundwater, and wetlands across the country. Only a few random studies refer to some of these negative consequences, such as:

1. Diminishing of groundwater discharge as a consequence of the retention of river water in the dams, the deviation of large water quantities from the river beds, the development of ballast quarries near groundwater wells;

2. Drying up and deterioration of the state forests and pasture lands that are located below collection channels for dams;

3. Loss of biodiversity, specifically among fish populations, in the rivers;

4. Deterioration of some alpine lakes from explosions related to the construction of hydro-engineering tunnels;

5. Drying of wetland zones as a result of reduced influx or direct, excessive exploitation.

In Bulgaria, wastewater has limited applications because of the underdeveloped state of sewerage systems (they are accessible to only 68% of the population) and urban wastewater treatment plants (available to only 36% of the population). The absence of incentives for wastewater treatment and reuse, the inadequate technical characteristics of the existing irrigation systems for utilization of treated wastewater, and other reasons are responsible for the limited use of wastewater.

The aggregate average annual amount of used water during drought is estimated at $16km^3$ and varies between 23.5 and $10.7km^3$, depending on the humidity of the specific year. Together with the water of the Danube River, used for Kozlodui NPP, water use amounts to $27.5km^3$ in humid years and $14.7km^3$ in dry years (Table 18.2).

Water Use

Water demands of the major water-users and their changes throughout the period 1983-1994 may be obtained from the data in Table 18.1.

Drinking water supply: The number of human settlements that rely on water supply systems is very high. Water is provided to 82% of the human settlements (100% of the cities and 81% of the villages), where 98% of the population lives. The capacity of the constructed water sources for human settlements is estimated at about $2.2km^3$ in dry years, whereby 72% of it is provided from groundwater, 9% from river watersheds and 21% from regulated dam water. The maximum amounts of drinking water used during the period 1982-1990 vary between the range of $1.65-1.73km^3$. The main reason for the lower performance characteristics of the potable water sources is thought to their diminished discharge rate "due to the construction of hydro-engineering facilities in the rivers, such as dams, derivations etc., that check and deviate the water from one river valley to another and deteriorate the hydrologic and hydro-geologic conditions" (Committee for Spatial and Urban Development, 1989).

Since 1992, consumption of drinking water has dropped sharply down to 1.2-1.3km³, both because of the current general economic depression and because of the sharp rise in the price of water services. The capacity of water sources has meanwhile increased because two irrigation dams with a total useful capacity of 0.27km³ have been re-designated to provide water for human settlements. Nevertheless, about 6% of the cities and 2% of the villages live under year-round water rationing, while 43% of the cities and 21% of the villages suffer seasonal water rationing. Table 18.3 shows the inadequacies of public water supplies. It includes national and city water loss amounts and water consumption by consumers.

The analysis of the data together with results of additional studies carried out in the period 1992-1997 reveal these conclusions (Dimitrov 1998):

1. Water losses during the period 1982-1994 increased 2.6 times, from 17% to 46.5%, and by 1995 exceeded 50%;

2. The high rate of losses is due to the depreciated water supply network (80% of which feature asbestos cement pipelines), as well as to the poor management of the systems (e.g., overflowing of urban reservoirs, increased rate of leakage due to insufficient pressure regulation, water thefts, and absence of water metering devices);

3. The use of water by public water supply systems for other purposes (industry and other types of consumers) is quite high and often exceeds the rate of water consumption by households. This is most typical for the cities with the gravest water rationing regimes, Gabrovo, Vratsa, Lovech;

4. In the cities of Pernik, Blagoevgrad and Vratsa the water rationing regime does not produce positive results because the population stores large quantities and hence consumes larger quantities of water;

5. In certain human settlements the available capacity of urban reservoirs is insufficient to equalize the daily (24 hours) and weekly differences between water needs and supply;

6. Some dams used for drinking water supplies have hydro power plants built below them (Studena, Iskar, Smirnenski) and are operated under an energy generation duty cycle as well. That is one of the reasons for the water rationing regimes implemented in the adjacent human settlements (Pernik, Sofia, Gabrovo).

Industrial water supply: The rapid industrialization of the country after 1950 led to sharp increases in water consumption by the industry sector. For that reason, a number of measures were undertaken in the period 1975-1985 to

Table 18.3 Household water supply in Bulgaria, 1995

№	Place	Water supply, $10^6 m^3$	Losses, %	Total water used, $10^6 m^3$	Household supply $10^6 m^3$	l/cap/d
1	Sofia	207569	57.2	88926	53362	131
2	Varna	63649	61.8	24297	10733	97
3	Plovdiv	46980	47.0	24879	12933	103
4	Burgas	39145	37.6	24412	9383	129
5	Russe	35124	49.0	17913	7283	118
6	Pernik	32698	31.2	22482	4043	122
7	Stara Zagora	29958	57.4	12777	4847	88
8	Gabrovo	27903	71.2	8028	1834	85
9	Pleven	26534	51.8	12802	5468	117
10	Shumen	23632	54.9	10649	3077	87
11	Sliven	19862	50.8	9770	3812	97
12	Veliko Turnovo	15065	49.2	7650	2462	99
13	Yambol	14080	64.0	5062	2919	89
14	Dobrich	13322	67.5	4331	2387	63
15	Targovishte	12000	73.4	3186	1154	74
16	Haskovo	11517	42.3	6644	2837	96
17	Vratsa	10682	48.1	5540	3788	135
18	Devnya	10247	31.2	7052	316	97
19	Blagoevgrad	9580	29.3	6768	3267	122
20	Lovech	9537	64.9	3346	1447	83
21	Remaining cities	289598	44.3	161102	85385	
22	Villages	246646	49.8	123841	88303	
22	National total	1201371	50.8	591457	305636	137

Source: The table has been compiled from information published in the Public Works and Communal Sector Bulletin of the National Statistical Institute. After 1996 the publication of the bulletin has stopped and the information contained in it has been defined as "confidential", i.e. inaccessible for analysis.

reduce water consumption in many industrial enterprises. Measures included water recycling, limiting the consumption rate of potable water, especially from the drinking water supplies of human settlements, and the construction of proprietary water sources. By 1983 the total amount of industry water consumption amounted to 2.09km³, 60% of which came from proprietary facilities, 28% from water supply companies in human settlements, and about 12% from complex dams (Committee of Spatial and Urban Development 1986). In the end, only 25% of the industry's water use was reduced during the transition period and economic depression after 1990, although the drop in production output in industry exceeded 50%.

In the past 15 years no studies have been conducted on water losses, efficiency of water use, or adherence to rated norms for water supply in the industrial sector. Data about water consumption in industry has a low degree of reliability. This is because the majority of the industrial water pipelines have malfunctioning (or no) water metering devices.

Irrigation: The existing network of irrigated areas in-situ at the beginning of the drought period is estimated at 1.19×10^6 ha. Forty-one percent of the areas were gravity supplied and the remaining 59% were supplied from pumping stations. Water supplies for irrigation were characterized by huge losses because many were open canals and the predominant irrigation technologies were designed for surface irrigation (on 50% of the irrigated areas) and spraying (on 49% of the areas). The distribution of irrigated areas by types of water sources is as follows: 53% from regulated water, 44% from running water, and 3% from groundwater. Water demand for irrigation varied from $3.4 \mathrm{km}^3$ in an average year to $5.26 \mathrm{km}^3$ in an average dry year. The normative probability of irrigation water is 75% in Bulgaria. In order to provide water for irrigation, 1,847 small water reservoirs with a total capacity of $0.553 \mathrm{km}^3$, 166 large dams with a total capacity of $2.527 \mathrm{km}^3$, and numerous intakes and other facilities were constructed to make use of $2.2 \mathrm{km}^3$ of inland river runoff. Supplies of about $1.7 \mathrm{km}^3$ of water from dams of complex purposes with multi-annual runoff regulation was envisaged for medium-dry and dry years.

In 1991, land reform was launched in Bulgaria. Its goals were to abolish the state institutions operating agricultural co-operatives and agro-industrial complexes and to reinstate land ownership rights of the former owners. Restitution resulted in parcelization of the arable land which delayed the implementation of the reform. Coupled with the general economic depression and the loss of markets, it also led to drastic diminishing of the irrigated areas and consequently, water consumption for irrigation. For the period 1991-1994 alone the irrigated areas diminished from 0.24×10^6 ha to 0.09×10^6 ha and water consumption for irrigation diminished from $1.2 \mathrm{km}^3$ to $0.24 \mathrm{km}^3$. For that reason a large number of the dams and other facilities for irrigation remained but had no consumers.

The latest reassessments of the irrigated land stock indicate that by 1994 the land area that was technically suitable for irrigation amounted to 0.6×10^6 ha, and the economically effective area amounted to 0.3×10^6 ha. The water demand for irrigation in the future is estimated to amount to about $1.5 \mathrm{km}^3$ (Ministry of Agriculture and Forests 1998).

Hydro power generation: After 1944 hydro power generation development was given priority, outpacing other water supply provisions in meeting the demands of consumers. During the drought period 1982 -1994, there were 91 HPPs with a total installed capacity of 2,398 MW and power output of 3,515 GWh. Only 9% of the installed capacity relied on running water and thus did not create serious conflicts with the interests of other water users. The remaining 91% were built below the dams and used regulated water. In this sense the water consumption of the hydro power generation sector is very similar to other water consumption sectors—it consumed more than 90% of its water requirements, especially in winter when energy consumption is the highest. The aggregated useful capacity of dams for power generation is 1.7km³ and that of complex dams, which supply the hydro power generation sector, is 2.5km³. The required water quantities for electricity output rates of the HPPs are estimated at a total of about 16.7km³, including 10.5km³ regulated water from the dams.

The comparative analysis with the parameters of the runoff of inland rivers reveals that the design parameters of the hydro power sector might be achieved only at a coefficient of runoff catchment and regulation above 82%, which is environmentally and economically unjustified for the prevailing conditions in Bulgaria (Table 18.2).

According to internal information from the National Electric Company, the operation of HPPs uses average annual water quantities of 10.5km³, 6km³ of which is regulated water; 6.4 and 5.7km³ of water were used in 1993 and 1994, respectively. These data differ considerably from the information provided by the National Statistical Institute, according to which the water processed by the HPPs was 2.8 and 2.9km³ in the same two years. The difference in the quoted data originates from the application of different methodologies, although intentional manipulation of the data with the aim to prove the necessity of building new dams is also possible. In actuality, water conflicts and water crises in Bulgaria are most frequently caused by unrealistic design parameters and the highly rated hydro power sector. The motivation of the energy sector to support hydro power is to retain its monopolistic position with respect to the national economy. Another main reason for the Bulgarian water crisis is non-compliance with the Water Act where water supply priorities are specified.

Generally speaking, the efficiency of the Bulgarian hydro power sector from the point of view of the water economy is contestable, since the generation of 1 kWh of electricity at a market price of US$0.035-0.04 is only possible by the utilization of about 3.5m³ of water. The population pays from

US$0.30 to 0.90 for 1m³ of water used for household needs, a price that varies in different regions of the country.

The hydro power sector does not pay a tax for consumed water quantities. The economic crisis after 1990 affected the energy sector to a lesser degree because it exports electricity to some neighboring countries. Therefore, the "new" energy strategy envisions a twofold increase in the capacities of the hydro power generation sector and large-scale construction of new dams and collection derivations, including transfer of the runoff of the transboundary rivers Struma and Mesta.

Balance of water demand and water supply In actual fact in Bulgaria there are no incentives for market-based relations in the water sector and for accurate identification of the parameters of water supply and demand, since:

1. Both water and water sector systems are entirely owned by the state. Private ownership rights exist only for some individual water intake facilities of industrial enterprises;
2. Water amounts drawn from the water sources is not metered but is rather determined by calculations based on the capacity of the water intake facilities and the number of hours in operation. In the case of the hydroelectric sector, they are determined by the amount of power generated;
3. There are no monitors or controls on the quantity of drawn water and no mechanisms to regulate and sanction excessive consumption and water waste;
4. There are no charges for "raw water," for drawn water quantities, or for runoff transfer, storage, or regulation;
5. The price of water covers only part of the costs for its extraction, transportation, distribution and delivery to the consumers;
6. Water management is performed under the rules of the centralized planning economy and has not been updated since the water sector systems were designed;
7. Structures that govern the water use and provision of water services are monopolistic and without market-based regulators.

Taking into account the above comments, Table 18.1 illustrates the changes in water use by economic sectors and by water sources in the period of drought 1983-1994. The analysis of the available information is made difficult by the combination of two factors influencing water consumption, the drought and the changes in the economic circumstances during the period under review. For instance, 1983 as a representative year was characterized by advanced economic development—there was a five percent increase in the Gross Domestic Product (GDP)—while it was a medium-dry year the water economy could meet water demand. The GDP growth in 1989 was only 1%. The years after 1991 are characterized by a huge drop in the economy by 30 to 50%.

Generally speaking, the data about water use follows the trends of the economic situation and indicate a drop in water consumption in 1994. In 1994, water consumption was 46% lower than it was in 1983. The share of water use by hydroelectric generation is the largest and relatively constant. In absolute terms, water consumed by hydroelectricity was negligibly reduced in 1993 and 1994. This is probably a consequence of drought, although it is possible that part of the reduction may be due to the economic crisis and the drop in energy consumption. In terms of its value relative to the general water consumption, the share of consumption by the hydroelectric sector increased from 34% in 1983 to 49% in 1994.

The largest reduction in water consumption was in irrigation, a 14-fold decrease in the investigated period. The reasons for this reduction are definitely related to land reform rather than to the drought.

Water consumption by households was relatively constant at about 0.36km^3 and accounted for an insignificant portion of total water consumption. Meeting this demand in the future should not be a problem even during the driest year or a series of dry years, provided potable water supply is given a priority.

The data reveal great reductions in the use of water from the Danube River (five-fold decrease after 1991) because of the introduction of regionally differentiated prices of water services and the high delivery price of water from the Danube. Until 1991 the price of drinking water in Bulgaria was the same for the entire country. The reduction in the use of water of the inland rivers is probably the result of both the drought and the drop in water consumption because of the economic recession.

Table 18.2 attempts to show the degree with which available water resources met the demands of consumers during years of differing humidity rates. The table was developed under the following assumptions:

1.   Water demand for households was determined at the normal rate of 180 liters/capita/day and population of 8 million inhabitants;
2.   Water demand for the services sector (e.g., trade and tourism) was determined from data for hotel accommodation capacity and its utilization at rates corresponding to the population size;

3. Water losses were assumed to be 20% of the total water consumption by the households and the services sector;

4. Water demand for agriculture included two components. Water demand for irrigation was calculated on the basis of the established effective irrigated areas (0.3 million ha) at a rate of 3,000m³/ha in a medium-humid year, 4,000m³/ha in a medium-dry year and 5,000m³/ha in a dry year. Water demand for stockbreeding was assumed to be equal to the annual average records for the period 1982-1989 when relatively optimum development parameters were observed;

5. Water demand for industry was determined under the assumption that it would recover its 1985 levels. If it exceeded 1985 levels, it was assumed that this was accomplished by advanced technology where water use for technological needs retains its rate of about 2.1km³ per year;

6. In determining the ability of water resources to meet demands, present capacities of the water catchment facilities were taken into account. However, other management parameters were added. Parameters that were added include reduction of losses in water extraction and water transportation, increased capacity of the systems for recycling water supplies, frequency with which the available water resources were used, effective diminishment of the volumes of multi-annual regulation of runoff, increased capacity of the reservoirs for equalizing daily and weekly water consumption imbalances, and compliance with water use priorities specified in the Water Act.

Table 18.2 shows that the available water sources (less the Danube River water used for Kozlodui NPP) may secure an average annual amount of about 16km³ of water for different applications, which in dry years may drop to 10.7km³. About 4.1-4.7km³ are necessary to meet the water demands of industry, agriculture, households and the services sector. Available water resources may adequately provide this amount. The amount of water that will probably be consumed by the HPPs amount to about 11.4km³ in an average year and diminish to 6km³ in dry years. This indicates that the hydro-power sector should strive to reach its design capacity not in terms of water consumption, but rather with respect to its electricity output alone. This should be done by improving the coefficient of efficiency of the units and installing new capacities in the dams that have remained unused (e.g., Ogosta, Pchelina, Sopot, Krapets, Dolni Dubnik and others with a total storage capacity above 0.9km³) and in other water supply facilities as well.

Management Structures and Practices in the 1982-1994 Drought

Keeping in mind the limited water resources of the country (annual average is about 2,400m³ per capita), the high frequency of dry periods, and the large scale of the in-situ infrastructure, it is interesting to look at the organization of water management itself in terms of management structures, mechanisms of interaction and tools of impact and regulation of the water sector. Generally speaking, water management in Bulgaria may be characterized as "centralized decentralization" or "concentration and decentralization at the national level." This means that the responsibility for different aspects of water management is assigned to numerous central institutions. More detailed information of the institutions and their specific responsibilities with respect to water management may be found in Table 18.4.

Table 18.4 shows that four ministries manage water use through their planning, design, investment, and operation structures without any co-ordination among each other. The ministries and their primary functions are as follows: the Ministry of Regional Development and Construction, with respect to the drinking water supply of human settlements; the Ministry of Agriculture and Forests, with respect to the irrigation of arable land; the Committee of Energy, concerning the use of water for energy generation purposes; the Ministry of Industry, concerning the use of water for industrial purposes. There are other numerous institutions responsible for monitoring the quantity of water resources, the Bulgarian Academy of Sciences, the Committee on Geology and Mineral Resources, the National Electric Company, the Ministry of Transport, and others.

In 1985 the National Water Council was created with the aim to coordinate and balance the interests of all stakeholders. Originally it was a part of the Ministry of Regional Development and Construction, later (after 1991) it became an advisory body to the Council of Ministers. The National Water Council consisted of a Chairman and representatives of the Ministry of Regional Development and Construction, the Ministry of Agriculture and Irrigation Systems, the Ministry of Industry, the Ministry of Transport, the Committee of Energy and the National Electric Company, the Ministry of Health Care, the Ministry of Environment, the Civil Defense and the Committee on Geology and Mineral Resources. The principal responsibilities of the National Water Council were related to the management of water quantities and water uses. In more detail, responsibilities included the issuing of water use permits; establishing monthly, quarterly and annual limits for water extraction from the largest and most important dams; and forecasting future water demands and design measures to meet them. The National Water

Table 18.4 Institutions and their water responsibilities in 1982-1995

Institutions	Responsibilities
Council of Ministers	Decision-making for construction of large hydro-engineering facilities. Fixing of the pricing and tariff policy.
National Water Council 6 regional administrations on the basin principle	Issues permits for water use. Develops monthly, quarterly and annual duty cycle schedules for drawing water from 27 complex and indispensable dams. Sets monthly and annual limits for water use for different applications from the 27 dams.
Ministry of Regional Development and Construction Water Sector Division Regional Development Division Local Authorities Division	Application of the state policy on water supply and sewerage in human settlements; elaboration of ordinances for use of the water supply and sewerage systems and methodologies for formation of the price of drinking water; methodological guidance to the regional and local authorities.
Vodokanalengineering	Design of water supply and sewerage systems in human settlements; elaboration of projects for categorization of river courses.
Research and Development Institute on Water Supply and Sewerage	Research and development activities in the field of water supply, sewerage and water treatment.
National Institute of Regional Development and Housing Policy	Elaboration of spatial development plans by municipalities and regional development plans by districts, including on water use and water protection.
Vodokanalinvest	Distribution and servicing of state water supply and sewerage investments in settlements.
Hydrostroy	State-owned construction organization for hydro-engineering construction.
Districts	Responsible for implementation of state policy at the regional level.
Municipalities	They exercise ownership rights on the small dams (a total of 1847), develop municipal development plans, incl. on water usage.
Water Supply and Sewerage companies - 29 state-owned and 14 municipal property	Operation, maintenance and repair of the water supply and sewerage systems (incl. 10 dams for drinking water supply), delivery and sale of drinking water to subscribers.

Ministry of Agriculture, Forests and the Agrarian Reform	Determines state policy in the field of hydro-melioration; proposes the amount of hydro-melioration charges and the price of water for irrigation.
Investment Policy Department	Determines the investment policy in the field of hydro-melioration, including irrigation, protection against flooding etc.
Irrigation Systems with Head Office and 26 regional structures	Operation and maintenance of the state-owned irrigation systems and dams (a total of 166), sale of water for irrigation, industry; water consumption, fish nurseries etc.
Agrovodinvest	Distributes and services state investments in the field of hydro-melioration.
Vodproekt	Design and planning of the development of hydro-melioration.
Institute of Hydro-engineering and Melioration	Research and development in the field of hydro-melioration.
Water Economics	Construction of hydro-melioration facilities.
Committee of Energy	Formulates and manages state policy in the field of energy and energy resources, including hydro-power generation.
National Electric Company	Heat and power generation, supply, distribution and sale.
Dams and Cascades Division with 6 regional divisions	Operation and maintenance of 12 dams, watershed and collection derivations, as well as the derivation for transfer of water between different river basins; monitoring of precipitation rate and the runoff of watersheds of the hydro-power dams; metering of water levels in the hydro-power dams and determination of the water influx and discharge rates from them.
Energoproekt	Design of hydro-power cascades and other facilities for utilization of the energy of water catchments, collection derivations and derivation for water transfer, HPPs.
Ministry of Environment and Water Protection of Water Cleanliness Department 16 regional inspectorates	Formulates state policy on environmental protection, including water; works out ordinances, plans, programs etc.; monitoring and control on environmental pollution, including water pollution.

National Center on the Environment and Sustainable Development	Monitoring the state of the environment, incl. waters; water sector cadaster; Bulletin on the state of the environment, incl. water quality.
Ministry of Health	Responsible for the quality of drinking water and waters for swimming, as well as for mineral waters.
28 hygiene and epidemiology inspectorates	Monitoring and control of the quality of drinking water and the coastal waters for swimming.
Ministry of Finance	Responsible for allocation of financing of sites from the state budget, including hydro-engineering facilities of national importance, small water supply facilities etc.
Ministry of Industry	Prepares justifications for water demand for industrial applications.
Ministry of Transport Administration for Maintenance of Navigation Along the River Danube	Monitors water levels and water quantities along the Bulgarian section of the river Danube.
Committee on Geology and Mineral Resources	Hydrogeologic investigations and monitoring of underground waters.
Civil Defense	Protection of the population in emergency cases–floods, calamities, breakdowns etc.
Bulgarian Academy of Sciences	
National Institute of Meteorology and Hydrology	Collection and processing of hydro-meteorological information. Monitoring of the levels of underground waters and the water quantities in inland rivers.
Institute of Water Problems	Research and development activities in the field of hydro-engineering construction, safety of dam walls and water sector studies and analyses, water quality and treatment, etc.
Institute of Geology	Research and development activity in the field of geology and hydrogeology.
Forest Research Institute	Research and development activities in the field of forest hydrology and the impact of forest management on the water balance of watersheds, climate change impacts etc.

Council met once a month. Its activities were supported by the services of its Head Office and six Regional Basin Administrations.

It is interesting to review the organization and the application methods of the two principal management tools used within the framework of the National Water Council. The permits for water use were issued upon submission of an application and a hydrologic or hydrogeologic report (depending on whether the application is for the use of surface waters or groundwater). The report could be approved by the Experts of the National Water Council. The quality of the hydrologic reports was often low because of the poor quality of accessible information and because of the absence of standards for the reports. No "environmental impact assessment" was required for the issue of permits for water use. There was no complete register of permits already issued because they had been issued during different periods by different institutions and had not been systematized by river stream sections or by hydrogeologic regions. In this way some of the permits were issued without guarantee that the permitted quantity of water resources was available. There were also no data in the permits about the exact location of the water sources and the facilities that would be used for water extraction, no guidelines for water users' responsibility to ensure water influx to riverbeds, nor were there proper definitions of the rules of behavior under conditions of drought.

The system of management of the water supply dams oversaw 27 complex and important dams with a total storage capacity of 4.9km³ and a net capacity of 4.2km³. Six of the total number of dams were designated for drinking water and 21 had complex purposes. The structures responsible for their operation and maintenance included: nine dams were put under the responsibility of the water and sewerage companies of the cities of Sofia, Burgas, Veliko Turnovo, Gabrovo, Vratsa, Sliven, Kardjali and Pernik. Six of them were operated by the Irrigation Systems and its regional structures in Pernik, Sliven, Pazardzhik, Varna and Sofia, and 12 dams were operated by the Dams and Cascades Division of the National Electric Company.

For management of the dams, experts on the National Water Council worked out the so-called "duty cycle planning schedules." They were compiled on the basis of the applications of water users and information supplied every month by the operating institutions of the respective dams. These institutions supplied data about the available water volumes in the dams and the influx and discharge of water from them during the preceding period. Out of all of the submitted information only the water level was actually metered. The assessment of the available water volume in the dam was done on the basis of the so-called "key curve," i.e. the graphic relationship between the water level and the water volume in the dam, established during the project design. The rate of consumption was assessed according to the capacity of the dam and the operating hours of the water extracting equipment, most of which were pumping plants or HPPs. The income was calculated according to the balance sheet of the discharge rate and the difference in the water volumes in the dam of the preceding and current periods. There were three types of duty cycle planning schedules, those for average, dry and humid years. Schedules included options for the issue of permits for quantified limits of different water users. These schedules were viewed as typical because no count was taken of the entire range of probable monthly values at identical availabilities of the annual runoff. Schedules were usually drawn up using the baseline method. Most frequently, limits set in planning schedules were based on other years in the dam balance records that had similar runoff parameters.

The planning schedules were then submitted to the National Water Council for review and approval of the proposed limits. At the end of each year, duty cycle planning schedules were developed for the next year. They were subject to monthly adjustments depending on the changes that occurred in the rate of income, the rate of consumption and the available volumes. There were no structures or systems for controlling the implementation of the allowed limits and sanctioning of violations.

The planning documents in the field of water resources were drawn up by municipalities, districts and for the entire country. Usually they represented a mechanical sum of the visions and proposals of the institutions that were responsible for the development of the individual economic sectors with respect to the various water uses. No market-based approaches or river basin methods were applied in the process of planning. For that reason, the methods used were not precise and errors were made in the hydrological, economic and environmental aspects of water management. The planning documents were developed uniformly by teams of hydro-engineers. Therefore, there was no integrated or interdisciplinary approach even though the proposed measures were normally related to programs for new construction in the first place.

The scheme described above reveals some very significant omissions and shortcomings of water resources management in Bulgaria, namely:

1. Lack of clarity, accuracy and precision in the regulation of the issue of permits for water use;

2. Lack of economic mechanisms and market regulators in both the planning and construction of water sector systems, as well as in water resources management;
3. Permanent administrative intervention in water affairs;
4. The use of information that was not sufficiently comprehensive and featured a low degree of reliability;
5. Inadequate expertise of the decision-makers and absence of feedback available to them;
6. Absence of modern devices for monitoring incoming and outgoing reservoir water;
7. Absence of water meters and absence of a system for monitoring and controlling water consumption;
8. Failure to apply modern and scientifically-based methods for forecasting runoff, assessing real water demands, and optimizing water extraction from the dams.

The 1994-1995 Sofia Water Supply–A Crisis Caused by Drought or Mismanagement?

All of the above-described deficiencies of water resources management found their greatest manifestation in the crisis of the Sofia City water supply in the period 1994-1995. The population of Sofia increased almost threefold in the past 50 years. From about 400 thousand inhabitants in 1945 it grew to 1.34 million by 1972 and then again dropped to 1.19 million by 1993. Two large industrial zones were created as satellite appendices of the city, and a 15×10^3ha vegetable-growing belt was created, supplied by an irrigation system. The water supply system of the city developed parallel to the development of the city itself. It is comprised of the following supply and treatment elements (Figure 18.1):

1. The Boyana water supply pipeline, whose capacity is about $2 \times 10^6 m^3$ per year. The water is fed to the water pipeline after its treatment in HPP Boyana, dimensioned for $3.9 \times 10^6 m^3$;
2. The Rila water supply pipeline, whose throughput capacity is $2 m^3/sec$ or above $50 \times 10^6 m^3$ annually. Initially, the pipeline was fed by river intakes on tributaries of the Iskar River (Levi, Cherni and Beli Iskar). In 1945 the Beli Iskar dam was added to the system (total capacity $15.3 \times 10^6 m^3$, including $14.9 \times 10^6 m^3$ useful capacity). In 1982 the dam was connected to the Belmeken-Sestrimo cascade by means of a reverse tunnel. This created an opportunity to re-distribute the water from the Struma and

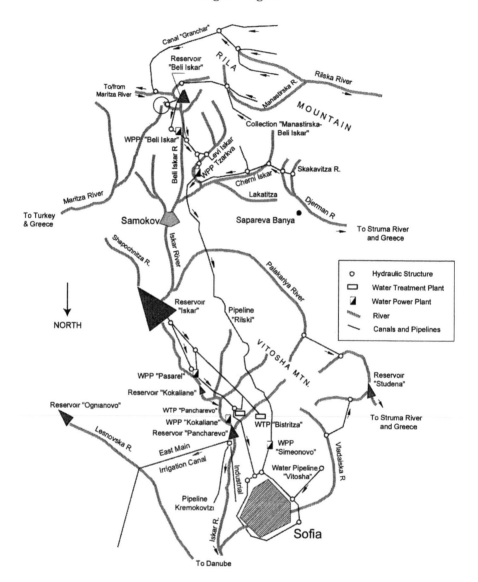

Figure 18.1 Water supply of Sofia

Mesta Rivers to the Maritsa and Iskar Rivers. After 1983 transfer of water to the Beli Iskar dam began and in some years it exceeded $100 \times 10^6 m^3$ (e.g., 1986, 1987, 1991). The water from the Rila pipeline is treated by the HPPs Beli Iskar, Mala Tsarkva and Simeonovo, whose design capacity is respectively 54.8, 65.4 and $39.2 \times 10^6 m^3$;

3. The Iskar pipeline (fed by Iskar dam), which is fed from three points, from the high-pressure derivation above HPP Passarel, via the emergency connection along the Passarel pipeline after HPP Passarel, and along the peak-duty connection from the high-pressure tunnel above HPP Kokaliane. The water from this pipeline supplies Sofia after being treated in the drinking water treatment plant Pancharevo, whose capacity is 4.5-5.0m^3/sec. Usually a larger quantity of dam water is fed through the water pipeline and is mixed with quantities fed by the treatment plant;

4. The Iskar Dam, a multi-annual equalizer with a gross storage capacity of 673x10^{6}m^3, of which the net storage capacity is 580x10^{6}m^3. It is the principal source of water supply to Sofia, as well as the main element of the Iskar Cascade which supplies other water demands as well. Thanks to its large capacity, the dam can store, process and re-distribute both the income of its own watershed, estimated at about 330x10^{6}m^3 annual average, and the water transferred from other watersheds. The original designation of Iskar Dam was for irrigation and hydropower generation. That is why the HPPs Passarel and Kokaliane were constructed below it for processing of an annual average of 330 and 350x10^{6}m^3 of water, respectively. This is also why two equalizers, the Kokaliane dam wall, a daily equalizer whose capacity is about 2.7x10^{6}m^3, and the Pancherevo dam, a weekly equalizer of 6.7x10^{6}m^3 rated capacity, were constructed below it as well. Part of the water treated by the HPPs is fed via the Pancherevo Dam wall for irrigation and water supply to the industrial zones and the Kremikovtsi Metallurgical Plant;

5. The Ognyanovo Dam with a total capacity of 32x10^{6}m^3 was designated for irrigation and industrial needs and was constructed on a tributary of the Iskar River outside the watershed of Iskar Dam and its equalizers. It is not used because of the lack of commissioning protocol and an in-situ infrastructure for water use;

6. The water demand for households at a rating norm of 200 liters/capita/day is about 87x10^{6}m^3 net consumption. In the past, water consumption in industry amounted to about 100x10^{6}m^3. However, as a result of the measures undertaken (construction of proprietary water sources, revolving water supply etc.) it has dropped to about 25x10^{6}m^3. Water demand for irrigation is estimated at about 45-50x10^{6}m^3; however because of the urbanization of part of the agricultural land and decrease of irrigated areas, it has also dropped to about 15x10^{6}m^3 annual average. As a result of the economic crisis in the period after 1991 the demand for water by industry was also reduced. Thus the aggregate water demand for the city of Sofia, including losses of the amount of 50%, does not exceed 180x10^{6}m^3.

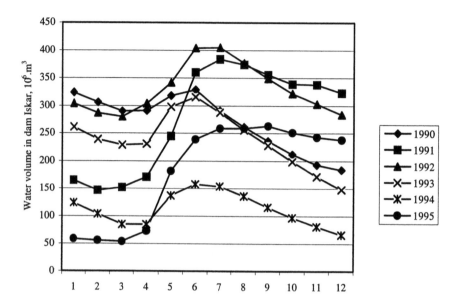

Figure 18.2 Monthly water volume in Iskar Dam, 1990-1995

The above review of the sources and demands for water in Sofia City shows that a powerful system of water sources exists around Sofia City that allows flexible management. For instance, there are supplies of large quantities of water from the river catchments during the spring high-water season and storage of large reserves in the dams for meeting the demand of dry periods. The capacity of the constructed water reservoirs exceeds the water demand for households, and no water supply crises should be expected for Sofia, even during a series of dry years, provided priority is given to potable water supply.

Nevertheless, it may be seen from Figure 18.2 that at the end of 1993 the water volume in Iskar Dam diminished to $148 \times 10^6 m^3$, and by March 1994 it dropped down to $80 \times 10^6 m^3$. This indicates that the threat of water crisis and a breakdown of the water supply to the capital were quite realistic. That development has triggered serious concern among water professionals and has stimulated public discussion for finding ways out of the crisis.

The first serious debate was organized at the beginning of April 1994 by the newly created Public Committee for Improvement of Water Supply to Sofia. This committee was lobbying for resumption of the construction work on the Water Supply Complex "Rila" and was trying to prove that the Sofia water supply crisis was due to the drought and the shortage of water sources. It was supported by the design institute "Energoproekt," designer of the Complex "Rila", and the University of Architecture, Civil Engineering and

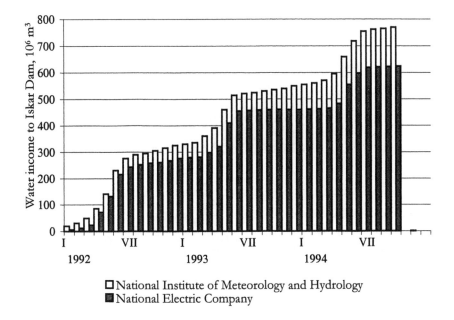

☐ National Institute of Meteorology and Hydrology
■ National Electric Company

Figure 18.3 Water income to Iskar Dam in the period 1992-1994, 10⁶m³

Geodesy, co-designer. The opponents were eleven institutes of the Bulgarian Academy of Sciences led by the Director of the Institute of Water Problems, NGOs (Ecoglasnost, Green Patrols etc.) and eminent public figures and experts. There the thesis was defended that water was not managed wisely and was wasted. This discussion was followed by many others, but no common position was reached among the arguing parties. The principal contested issues that were present in all discussions were as follows:

1. *Data about the water income to Iskar Dam.* As evident from Figure 18.3, there is disagreement between National Institute of Meteorology and Hydrology and "Dams and Cascades" Division of NEC (National Electricity Company). It is true that both organizations meter the incoming water of the dam. However, because of financial restrictions the National Institute of Hydrology and Meteorology has neither a sufficient number nor appropriate types of hydrometric stations to determine the income to the dam. In contrast, "Dams and Cascades" determines the income by calculations on the basis of the electricity output of HPP Passarel, results loaded with a high degree of inaccuracy;

2. *Data on water consumption from Iskar Dam.* The data in Figures 18.4(a) and 18.4(b) indicate clearly that great differences exist between the official information and archival data of the balance of the dam with respect to

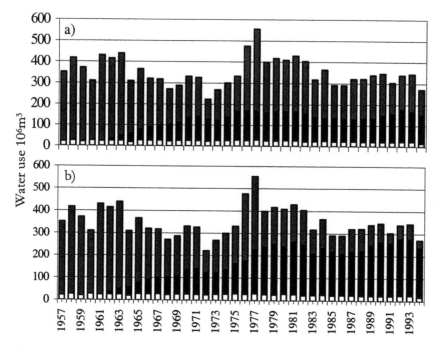

**Figure 18.4 Water use by sector from the Iskar Reservoir, 1957-1994;
(a) National Water Council data, (b) Official data. Striped = power,
irrigation and industry; solid = water supply**

the "water discharged for water supplies." The differences are probably due to an attempt to conceal the quantity of water drawn for the energy schedule of the two HPPs below the dam and its assignment to the category "water consumption by households." By this manner the thesis that new water sources for drinking water supply are necessary was supported, because in accord with the Water Law the state is obliged to ensure priority of drinking water supply to the settlements;

3. *Causes of the water supply crisis and the methods for overcoming it in the future.* Due to the concealment and manipulation of information the disputes continue. The crisis helped solely to identify the fact that water losses exceed 50%. However, the implementation of measures to reduce the losses started only after the Sofia Water Supply System was transferred in 2000 to a foreign investor, International Water. Despite the institutional changes in water management after 1997, nothing has changed in the principles and approaches for managing the dams. Therefore the risks of a new water crisis are quite real. Moreover, it may be said almost certainly that the next dry period will probably be used as an excuse by the strong construction and energy lobbyists in Bulgaria, who already have

agreements with construction contractors from abroad, to demand the construction of the Water Supply Complex Rila with funding from the budget.

However, an objective, unbiased, and professional analysis of the situation reveals that the causes of the water supply crisis in Sofia overlap with the causes for the poor management of national water resources, which were detailed in the previous sections. These causes are related to the low quality of available information that obstruct efforts to identify the real problems and to propose measures for their solution; the lack of expertise and the carelessness of the water administration; the priority in meeting the water demands of the energy sector over meeting drinking water needs of households, which is promulgated by the Water Act; failure to apply scientific methods and to evaluate the cost-effectiveness and market-based approaches in the management of water resources; the lack of control on water use and on compliance with the allowed consumption limits; the absence of mechanisms for encouraging public participation in water management; and shortcomings of the regulatory framework.

Proposals for Upgrading Water Resources Management

It is obvious that the management of water resources in Bulgaria is undergoing a deep crisis. It suffers from numerous shortcomings and needs a radical change in its philosophy, principles and approaches.

First, it is necessary to dissociate the functions of the regulator from those of the operator. The role of the *regulator* should be to limit water use through the issue of permits, economic regulations, and the sanctioning of violations. *Permits* for water use should have the same form and stature of a contract, in which the rights and obligations of the operator are outlined as well as his/her responsibilities and the punishments for rule violations. They should be detailed, precise and exhaustive, taking into account the realistic capacity of the water source, environmental norms, and the rights of the other water users, both individuals and landowners. *Charges* for water use should play a regulating and sanctioning role in order to compel the operator to save water, reduce water losses, and to build facilities only for water amounts for which there is real need and market. In addition, they should comply with the particular characteristics of the water source, for example its recovery capacity (groundwater recovers more slowly), vulnerability etc. Quite probably, their value should be fixed on the basis of the assessment of the environmental damage and the economic benefits of their use.

The *operator* should be allowed to independently manage the water sector system and assign water resources by virtue of the permit and in compliance with the enforced legislation. He or she should be allowed to do the job without permanent administrative intervention, but should be subject to regular control and sanctions in case of violation. When the operator provides water supply services, his or her relations with the subscribers should be based on contracts and compliance with the contracts should be controlled by the regulatory body. However, sanctions for non-compliance may be imposed by a rule of court as well, including indemnification for damages.

Second, it is necessary to conduct a comprehensive re-assessment of the income and design parameters of the existing water sector systems with the goal of defining their optimum duty cycles that will serve as a realistic basis for updating and re-issuing water use permits. It is also appropriate to re-assess the technical state-of-repair and economic efficiency of the existing systems and to formulate future policies about their privatization, concession contracts, and the dismantling of facilities that are in a poor technical state, are no longer needed, or are environmentally unacceptable. Penetration of private sector and private investments in the water business is an important prerequisite for regulating the legal framework of water economy and for improving the efficiency of water resources utilization.

Third, it is urgently necessary to develop plans for river basins and sub-basins on the basis of the integrated water management principles, in full compliance with the EU Framework Water Directive. They should be designed by interdisciplinary teams, and should comply with all requirements of the Directive. They should include precise, competent, and correct assessments of environmental damages inflicted by large-scale hydro-engineering construction. They should also include recommendations for mitigation of the negative consequences of such construction. Future hydro-engineering projects should also be re-evaluated in light of new environmental and economic requirements.

Fourth, as a result of new studies and re-assessments, it is necessary to develop an entirely new legislative package that addresses all of the above mentioned aspects and establishes a new legal framework. The most urgent legislative amendments are related to the creation of a strong and competent regulatory body in the field of water. Precise environmental regulations for the protection of water resources and clear rules for water business are also needed.

Fifth, it is necessary to create a competent and independent body, empowered by strong authority rights (for instance, a Water Agency), that would implement a thorough reform in water management and of water industry and water services. To this end it will be necessary to collect and systematize in an appropriate way all available information; set up a system for monitoring and pricing water use; conduct a thorough re-assessment of the state of affairs in the sector in order to identify problems and to justify required changes; prepare a comprehensive package of laws (Water Act, Law on Water Industry, and necessary by-laws) and a system of economic regulators, guaranteeing water sector reform; develop plans for managing river basins and measures for improving water resources and their management; and establish the necessary agencies for basin water management and mechanisms for effective public participation in it.

Sixth, until the above activities are implemented, investment priorities should be oriented to reconstruction and upgrading of water supply systems with a goal of reducing water losses and improving household water supplies; construction of sewerage systems and water treatment plants for human settlements with a goal to improve the quality of water and encourage the re-use of treated wastewater; and reforestation and stabilization of watersheds, particularly of the large, complex and important dams, that will have a positive impact on water resources and will reduce sediment load into them. This will prolong the life span of the dams and slow down the decrease in their useful volumes.

The greatest challenge for the future development of water management in Bulgaria is to change social attitudes and education, to improve public awareness of the value of water, and to make water conservation an official priority of state policy.

References

Committee on Spatial and Urban Development. 1986. Генерална схема за комплексно използване и опазване на водните ресурси в НРБ. София.

Committee on Spatial and Urban Development. 1989. Програма за водоснабдяване на Н Р България до 2010 година. София.

Dimitrov, G. 1998. Технически възможности за намаляване преразхода на вода в сгради и водоснабдителни системи. *Водно дело*, 31-37.

Gerassimov, S., N. Nikolova, D. Davıdov. 2000. "Water Resources and Hazards," pp. 199-228 in M. Staneva, C. G. Knight, T. Hristov, D. Mishev (editors), *Global Change and Bulgaria*, Sofia: CIRA and NCCGC-BAS.

Hristov, T., S. Yancheva, N. Nikolova. 1995. Защо не трябва да се строи ВК "Рила"? *Списание на БАН* 1:75-80.

Ignatov, P., N. Nikolova, G. Gergov, I. Raev. 1988. Становище относно предложението на Йордан Стаменов, направено пред ЦК на БКП, 28.09.1988, БАН, София.

Knıght, C. G., S. Velev, M. Staneva. 1995. "The Emergıng Water Crisis ın Bulgaria," *GeoJournal*. 35(4):415-423.

Mandova K., V. Tsaneva, R. Nikolova. 1998. *The Experience of the 3 Years of Operation of the Pilot Basin Council for the River Yantra*, ETP-USAID, Sofia.

Ministry of Agriculture and Forests. 1998. *Strategy for Development of Irrigation Agriculture in Bulgaria*, Sofia.

Order of the Government Council. 1990. No 137/26.01.1990.

Sendov, B. 1990. Становище на БАН по проектите на водоснабдителния комплекс "Рила" и хидротехническия комплекс "Места." Доклад до Председателя на Народното събрание, БАН, София.

Stamenov, I. 1988. Изложение до Политбюро на Централния комитет на БКП. 28 септември 1988, pp. 16 (unpublished).

Chapter 19

Media and Political Attitudes

Georgi Fotev

This chapter concerns the media coverage of and the political response to the water shortage which culminated in the end of 1994 and in the first months of 1995. As we are interested specifically in media and political coverage it is possible to distinguish two sub-periods in the time period under discussion. The first stretches from the beginning of the drought until November 1989 when the totalitarian communist system collapsed and the Bulgarian society was in total crisis; the second from November 1989 when society began an unprecedented transformation in the media and in political life.

It is evident that this division into periods is not formal but is useful in our case. During the first period, the media system and political system were an inseparable part of the monolithic social world of a totalitarian society. After the collapse of the system during the second period, society became pluralistic. With political pluralism, the media was freed from state control and supervision. Freedom of speech and public expression developed until new legal parameters of public life were established.

Our empirical basis is centered on observations on media behavior (printed and electronic) and on society's political life during the study period. Essential data will be obtained from the content analysis of two Bulgarian dailies—*Trud* (Labor) and the *Zemya* (Earth). There are three reasons to choose the *Trud* newspaper. First, this is the only nationwide newspaper that has been published during the entire period of observation (before and after the changes); second, its readership is the highest in Bulgaria; and third, after 1989, the newspaper was declared politically independent and allowed publication of various opinions, including those of political antagonists.

The *Zemya* newspaper is another story. This newspaper is the successor of the *Kooperativno Selo* newspaper (Cooperative Village) which under the communist regime carried propaganda of the Communist Party's strategies for developing the agricultural sector. During the beginning of democratic changes, the *Kooperativno Selo* was renamed *Zemya* (Earth) and remained similar in ideological orientation to the Socialist Party—basically the former Communist Party with a new name. However, the more important reason for selecting *Zemya* is that it is, perhaps, the single nationwide newspaper that debates agricultural issues, a sphere that, understandably, is directly affected by the water resources of the country and by the water shortage in particular.

The content analysis of both newspapers covered from 1 June 1982 to 31 August 1995. We selected issues from the two dailies at an interval of three years. In this manner the years of 1982, 1985, 1988, 1991, 1994 and 1995 are included in the sample. The last two years are successive because the peak of the drought occurred then. Within the framework of each study year we have selected the time interval 1 June-31 August; within this time interval we used every tenth newspaper issue. This scheme was selected to bypass Sundays, when the newspapers, as a rule, were not published. The issues in the sample were analyzed according to specific systematized indicators and subject lines. For example: we recorded page numbers that contained drought information; which region each article about drought pertains to (e.g., the whole country, a separate rural region, a city, etc.); what its temporal orientation is (past, present, future); what activities and spheres of the drought situation are discussed; the length of the article; the subject of study; author's status; specific features of the publication (is it describing the events, or prescribing solutions, is it politically biased (i.e., what is the political orientation); the types of conflicts that are discussed; and the purpose (to inform, to manipulate, etc.). Because the spontaneous and rapid events with respect to the implementation of the Djerman-Skakavitsa Project and the scandalous water regime in the country capital occurred during 12 November 1994-21 March 1995, all issues of the *Trud* daily during this time were analyzed separately.

No doubt, we could have had an even more complete picture of the data and representation would be of a higher degree if there were a systematic monitoring of the entire Bulgarian press and the electronic media, especially in regard to the issues concerned. We are not aware of such studies having been done for the drought period in Bulgaria; media studies and political analysis have been dedicated to other issues. Unfortunately, owing to a shortage of time and financial resources, it was not possible to include broader empirical data in our study.

It is necessary to consider a crucial methodological point. The key importance of the media in affecting public opinion in democratic circles is beyond dispute. A theoretical study of public opinion is of special importance for further analysis and is used as a basic methodological reference point. Public opinion is a part of social reality. It could be said that it is a phenomenon of a rather mysterious nature. It is not a realm based solely on truth but a mixture of truth, semi-truth, and delusions that are visible only at a distance and from scientific point of view (and to the scientific eye). Whatever the particular case, public opinions are extremely important because they can lead to social action.

Studying public opinion via mass media, we discover what the attitudes of society are and what different social groups support or refute, or are uncertain about. All this gives us information how public opinion becomes part of social reality even though it, by itself, is entirely virtual. Public opinion may ignore real consequences of drought or it may sufficiently influence the political response to drought.

In all modern societies, including Bulgarian society, the media plays an important role in public opinion formation, its management, and not uncommonly in its manipulation. It is very important for society's agenda to be represented in the media. When it is not represented, one gets the impression that something crucial and substantial is absent. A well-informed citizen is one thing and a citizen who is partially informed or not informed by the media is something completely different. A very significant issue is how the media highlights inconsequential aspects of an important problem and downplays key aspects.

Media Under Control and Society Without Public Opinion

The content analysis of the dailies makes it obvious that the frequency with which publications address effects of drought was higher during the Communist regime in Bulgaria. However, the question concerning the media style in regard to the drought process and the water shortage and also the political attitudes towards these problems is of greater importance. A particular kind of liberalization occurred in the system of Bulgarian media in the 1980s, especially in the middle of the decade. This occurred in the second part of the 1980s under the influence of the *perestroika,* which was instituted by Gorbatchov. However, the channels for mass communication in Bulgaria still remained under the control of the Communist Party's top clique. The media did not keep silent on matters concerning drought, but the usage of the "crisis" term was generally avoided, and in the media and other public spaces it was completely prohibited. The tangible facts as well as summarized data of the consequences of the drought and the condition of water resources were, in practice, kept secret and only a very small circle of experts and members of the Party and the state (i.e. the party-state) had access to them.

However, the phenomenon of drought was regularly explored as a justification for the failures in the system and explanations were given that cannot at all be perceived as its true results. On the contrary, predominantly through the media, political factors worked to create in the public mind a perception that the party-state had done everything in its power to remedy negative consequences of the drought. However, they usually reasoned that natural disasters are natural disasters and nothing more can be done. When public criticism arose concerning the lack of preparedness in meeting unfavorable natural phenomenon, critics were directed to so-called "small truths" (insignificant problems, causes and reasons) about the way the country developed and socialist society advanced. It is unbearable to think about any sort of criticism against the policy of the party-state in respect to the water issues which this country faced and about a public defense of opinions to the methods of approach and objectives of the central authorities. Naturally, a shortage of water availability to meet public needs because of polluted rivers and water basins, as well as many other negative domino effects from drought, cannot be hidden from the public eye. The concrete statements related to the problem in this regard are also concealed from society or, at best, the data were obscured, reduced or "softened" because of political and ideological speculative purposes. By doing so, the media propaganda was oriented towards mobilization of social resources for conservative and economical water usage. Of course, we cannot expect the effect from the manipulative propaganda to be tangible and clearly specified. Thus, an internal lack of interest and indifference was created and, if necessary, had to be kept hidden for the outsider by obscuring the reality through actively staging showy events.

It was natural that the rationality and irrationality of the established water management structures came to surface during the drought and times of water shortages. This became a key social and political issue. In the most arid regions with poor water resources, water regimes were enforced that were extremely hard on households. In such cases, the real state of irrigation equipment and networks became obvious. However, these key issues remained out of the political debate; there was no open media discussion about them. Thus, the public was not aware of them. Therefore, public opinions could not exercise power in the decision-making process. However, during the totalitarian system, such a common public opinion is identified with the publicly valid mythologies of the Communist Party ideology in power. Such public opinion I define as *quasi-public opinion*. In this sense, we may say it was a society without a public opinion. Nevertheless, one cannot deny the fact that empirical studies of public opinion had been conducted in the 1980s by the Central Committee of the Bulgarian Communist Party, where a specialized center even existed, but the results of this research were available only for the party-state nomenclatura. We have had no chance to analyze such data in respect to the drought. However, this is not very important in our case because we are considering only facts located in the public media and political spaces during the time under observation. The key factor is this: society and social groups could not freely use the public media to express their judgments and opinions about the water shortage and its widespread social impacts.

Naturally, society neither was nor could be indifferent to the water shortage, nor could government-directed opinions hide the multiple dimensions of the impact of water shortage. Experts and journalists tried to write in the media–indirectly and very discreetly using Aesopian language–about such items that those in political power, the totalitarian party, would have preferred to keep invisible for the broader public. It was no accident that the public protest in the very heart of the capital in the eve of the collapse of the Bulgarian totalitarian system was related to the water problem. It was a social explosion, like a spark setting fire to public discontent. However, this first open public protest was only the tip of the iceberg. The results from a study by the MBMD Sociological Agency are very significant. It was conducted on January 30, 1995, and is representative of Sofia society. The biggest portion of the capital's citizens, 86.6%, thought that consultation with expert opinion should be compulsory before starting to build any major project affecting the environment and human health. Conversely, about 1.7% of the citizens thought consulting expert opinion before taking action is not necessary. However, as we have already mentioned, publicity of such public opinion was impossible to instigate during the reign of the totalitarian regime. In fact, the specific interests and attitudes of society were concealed, and even by the media which was under control.

Media Orientations in Respect to Public Opinion

The space in the print media devoted to a certain text (article, comment, reporter's story, etc.) is of real importance as well. Our content analysis has shown that more than 50% of the materials concerning drought were published on the inner pages of both of the analyzed national daily newspapers during the time period of the study. Nearly 30% of the total quantity of the materials is on the first page. This demonstrates the significance of the problems concerning the issue and its public importance. The number of publications longer than 90 lines is comparatively high. They are followed by relatively short items with lengths of less than 30 lines. Medium-length stories were between 30-60 lines, and then came texts between 60 and 90 lines (i.e. relatively long stories).

The highest number of publications, approximately 50%, refer to rural areas. Stories concerning the country as a whole and the capital were equal in quantity. Stories about drought in cities (except Sofia) were the most infrequent. The genre of publications is also an indicator of the orientation to the present, the news information and reporters' stories about events prevail. Approximately 74% referred to the present, versus 4% to the past and 22% to the future; analytical comments that refer to the past or future are obviously secondary. In our view, the quantity of scientific and popular publications by scientists were not sufficient. Professional journalists and media writers were the prevailing component of all of the authors. Generally speaking, this is understandable and even natural. Keeping in mind the complexity of the issues connected with the water shortage and its importance at both national and global levels, we suppose that the media people may not have invited experts who could most competently discuss these multifaceted and vital subjects. In fact, the perception of the Bulgarian public is that drought issues are self-explanatory. In a complex society, this perception about such a phenomenon is deceptive, with a lack of real understanding of the drought.

Media interest was highest with respect to the consequences of the drought for agriculture and for household needs (53% and 37%, respectively). Naturally, the immediate impact of water shortage was felt in these circles and as a rule they are more sensitive to its effects. It could be considered worrisome that 50% of the publications did not have recommendations. In cases where we find them, recommendations are expressed in the form of appeals for water conservation and care for its purity. There were some debates about alternative water sources. However, we have come to the conclusion that they were not explicit and definitive, in general. Nearly one quarter of the texts included discussions on alternative irrigation techniques, which indicates public dissatisfaction with the existing irrigation systems and their technological insufficiency. The Bulgarian media, and the public opinion of which they are indicative, did not pay enough attention to industrial water usage (only 6% of publications). Had Bulgarian industry been developed with an eye on water efficiency, minimizing the pollution of water resources and purification of polluted water? Publications analyzing tangible drought consequences were very few. Media materials warning the public of certain events were more numerous, especially those publications that informed or interpreted certain problems or groups of problems with respect to the water shortage. Ecological issues commanded only 4% of publications.

The media sources under discussion first made an attempt to influence their readers through rationality (20%). Second, they rely on emotional influence (12%). The number of cases trying to shape positive attitudes towards decisions of the ruling authorities (11%) were twice that of the cases trying to shape negative attitudes (6%). As these are general data for the whole study period, it is clear that attempts to instigate negative attitudes on governmental decisions under the communist regime was impossible and thus were excluded from media sources. Only after democracy developed did the media begin to criticize government decisions. In fact, this became a leading trend. When we define the type of influence as rational or emotional, we have in mind the specific type of the predominant arguments and approach used in the articles. It is a matter of journalistic professionalism to shape rational attitudes to events and phenomena, on the one hand, and on the other, to motivate active practical pragmatic actions among the readers.

Drought and Water Shortage: The Conflict Areas

If the drought was considered a full-fledged natural disaster, it would evoke social solidarity with those impacted rather than lead to conflicts. However, as we have already pointed out, the lengthy drought and water shortages affect the interests of various groups in different ways. Thus, the sources of social tensions and conflicts, otherwise more or less hidden in normal situations, rise to the surface and result in negative consequences from irrational decision making or wrong management.

The small percentage of the drought-related publications paying attention to the conflict is disturbing; nearly 30% reported no conflicts. Some 12% reported conflict over government decisions, and conflicts over other issues such as pollution, food, or industrialization each appeared less than 5% of the time. Let us recall that the data reflects the whole period under observation. It is implied that the situation would reverse when the social and political system changed toward democracy. However, it is important for us to fully understand to what extent and on what grounds the conflicts concerning the drought and the water shortage are presented in the media. Modern sociology, in fact, does not perceive social conflicts as a one-dimensional negative phenomenon or as a social pathology. It is not rare for social conflict to act as the vehicle of improvement for society and as the instrument to find correct answers. So much the better if the media demonstrates sensitivity for the emerging social conflicts, in this case in regard to an enduring water shortage. In our case it is not possible to blame the media for the relatively insignificant presence of social conflicts in publications; due to political reasons, conflict in these areas would have been unthinkable during most of the time period under observation. It was impossible for the media, which was under strict control, to deal with latent social conflicts.

Social and institutional aspects of the conflicts in regard to the drought and water shortage are interesting. The greatest attention of the media was to conflicts between the attitudes about the water shortage among different groups (about 9%). Such conflicts were predominant during parts of the drought period, especially during the eve of the democratic changes in the Bulgarian society and immediately after. More of the media reports concern conflicts between citizens and the central authorities (6%) compared to their concern with conflicts at the local level with municipal authorities (2%). This is the case because the municipalities did not have enough authority or legally acknowledged competency to offer solutions to important water resource issues. To redistribute responsibilities among the central and local authorities and governing bodies is a forthcoming task in the democratic development process of Bulgarian society. Although environmental NGOs played an important part in the eve of the 1989 changes and the turbulent months and years afterward, they are still not strong enough to be a major influence on the water issues of this country (less than 1% representation). One of these organizations, however, played a crucial role during the unstable end of 1994 and the first months of 1995.

The Water Shortage as a Focus of Media and Political Life

Never before have drought and water shortage issues been so essential for the Bulgarian media as they were between mid-November 1994 and March 1995. There are two main reasons for this acute interest. A desperate, intense social conflict began and escalated among the population of the region where the Djerman-Skakavitsa Derivation Project was slated to be developed but was delayed for a considerable amount of time (see Chapter 20). This project aimed to redirect the water flowing from the Rila Mountains in order to meet the permanent water needs of the capital, Sofia. The government-appointed month for starting its construction was November 1994. The government's decision to start the project was based on its concern for the town of Sofia, suffering a severe water rationing crisis; its main sources of water, the Iskar and Pancharevo reservoirs, had catastrophically been depleted. How the water "escaped" from these dams is still a mystery, and some speculate that immense quantities were purposely drained (see Chapter 18). The social conflict is between the local citizens in the area of Djerman-Skakavitsa and the citizens of the capital who were in a disastrous situation. The conflict had political dimensions as well. Some of the political parties supported the construction of the derivation and others were against it. Experts were also split into two opposite camps. The Ecoglasnost Ecological Movement sided with the population protesting the building of the derivation. Keeping in mind the severe environmental consequences if the construction were to be completed, it is no surprise that the group took this position. When the Government decided to support the construction of the project, conflicts exploded between the local people and the central Government. Civic disobedience broke out until police forces took action. All this provoked social and political passions, and the situation soon became scandalous.

The second reason for the media boom in regards to the water shortage is related to the freedom of the press. The media in Bulgaria compete with each other in their search for techniques to catch the reader's attention. The content analysis of all issues of the national daily newspaper *Dneven Trud* (Daily Labor) shows that there were 152 publications during the study period mentioned above (90 issues total). This makes it clear that there was more than one publication per issue. This fact is rather important in itself. Almost every day there was material on the front page, which means that this was the leading news for the day. The result was that the media spread knowledge of the event into the broader national scale. This gave rise to an empirical sociological survey by the National Center for Public Opinion Studies and the Parliament. Table 19.1 shows results of this survey. Almost 26% of the adult population of the country thought this project should be constructed. However, nearly 35% was in favor of finding the middle ground. This group considered both interests, of the people living in Sapareva Banya, near the Djerman-Skakavitsa project, and of the capital citizens. A comparatively small portion of the Bulgarian population, in general, believed the project should be given up (8%). The number of the undecided people in this complicated situation was substantial (32%).

Table 19.1 shows the distribution of opinions in Bulgarian society with regards to gender, age, and educational level and with an eye to the preferred solution of the conflict situation provoked by the water shortage. the majority of both groups men and women agrees with the idea of compromise. There is also little difference among age groups, with the exception of the eldest people where number of respondents in favor of the project is highest. People with primary, high school, and more advanced educational degrees support the position of the Government.

Evidently the table does not demonstrate a high level of understanding of the event. Actually, the media coverage did not consist of clear, reasonable arguments either. However, a wide range of opinions was widely shown in the media and publicly declared. Here are some opinions of those who have proclaimed civic disobedience: "We are not against Sofianians (the citizens of Sofia), many of our relatives and close friends live there and they know about

Table 19.1 Public opinions about the Djerman-Skakavitsa derivation

Demographic Characteristics	Should the project be built?			
	Yes, in its present version	Not without a compromise	No, it should be abandoned	Uncertain or no reply
Gender				
Male	27.1	37.5	8.6	26.8
Female	24.2	32.1	7.2	36.5
Age				
Age 18-29	19.0	39.2	8.9	32.9
Age 30-39	20.7	39.6	7.7	32.1
Age 40-49	26.3	8.8	8.1	23.9
Age 50-59	31.6	5.9	5.9	26.0
Age 60 +	30.1	8.3	8.3	39.6
Education:				
Elementary	17.7	8.0	8.0	60.6
Grammar	22.0	8.5	8.5	40.5
Gymnasium	22.8	5.9	5.9	30.9
Vocational	27.3	6.6	6.6	22.0
College	17.2	9.5	9.5	22.4
University	45.5	12.2	12.2	12.1
Total	25.6	34.7	7.4	31.8

our situation. It is solely the authorities that do not have even the vaguest idea about it. And if they are speaking about any kind of regime, they should come here to see what it means to have water only one hour per day, and during the summer–only for 20 minutes." Publications included the death of fruit trees as a result of water regimes; there was even an announcement that a person died because of the water shortage. Some media comments asserted that the barricades near Sapareva Banya were put there for political reasons. Of course, there are sharp satiric writings, such as "On St. Jordan's Day we shall throw the cross in tanks" (St. Jordan's Day falls in mid-winter; people jump in pools or rivers to take out a cross thrown there). Some horrible stories also appeared: for example, nearly 1,400,000 citizens from 1,037 settlements in the country lived for years on end under a water regime and the comment was that they "like stoics, silently and quietly bear their own cross." Other publications reminded readers that the price of water in Bulgaria was among the lowest in Europe and called for a change in pricing. Questions concerning the leaks and wasted water were very acute. Being a common good, water was thought to be like sun and air. A lot of sharp criticism was published against the Government of the day and against all previous Governments for their senseless, inadequate water policies and the current difficulties of people as the result. In general, it was thought that the water shortage was mostly a consequence of shortsighted policy.

After the social and political storm ended in March 1995, the Bulgarian media and political elite of the country, and the society as well turned their attention to other matters of political and social life. The water problems disappeared from the pages of the newspapers and, in general, from the media and political scene. This was a sign that society, its political elite, and the media still did not perceive dealing with its country's water issues as a long-term project.

Conclusions

Owing to the democratic changes in Bulgarian society, the media in this country has become the "Fourth Estate." Informing and mobilizing civil society, the media balances the other three estates in forming public opinion and in exercising civil control against the water mismanagement and speculations of governing bodies. The freedom of the media is closely tied to its responsibility to accurately report societal water issues to its readers, to alarm them on time about critical situations, and to provoke debates before the water shortage occurred as a result of the drought. A political power that failed in its water policies should be sanctioned by its followers and the entire society through the mechanisms of democracy as untrustworthy and inadequate.

Keeping in mind that the media is free and independent, and that it is impossible to supervise, instruct, and verify, we recommend that, above all, the national intelligentsia (persons of letters, arts and sciences) should be more deeply involved in the water issues of this country and see the media's potential to exert influence over various social groups, political elites, and the entire society. Intelligentsia's and media's critical function is to bring out in the open what would otherwise be concealed. In order to be popular with the media, experts should use language understandable by a broader media audience; the manner in which they present their own concepts and opinions should attract the public.

Research centers and advanced academic agencies such as the Bulgarian Academy of Sciences, and the universities need specialized teams for communicating with the media. These teams should be skilled in gaining the media's attention with regard to the country's water issues so that the media captures public interest at every level–local, regional, district and national. The state and its governing bodies, as well as the academic and cultural institutions, nongovernmental organizations, and the entire society should be interested in gaining media coverage on water issues so that the reporting of these issues is simultaneously diverse and complex. They should invest in journalist training and in improving their debating skills in the field of water issues.

Academic journals and reviews should be written in a way that will be understood by the media. In connection with this study we have reviewed various publications and periodicals; they left us confused. We found insignificant topics, a heavy "academism," and total fragmentation of knowledge in combination with lack of concern about the applicability of the knowledge. It seems to us that, generally, the texts are addressed to a very small circle of readers. I do not want to be misunderstood. My worries are provoked by the lack of ambition for a broader view on the issues overcoming fragmentation and aiming at practical application of theory. If academics cannot work with the media, how can we expect the media to be analytical and competent with respect to water issues. Institutions such as the Bulgarian Academy of Sciences should publish attractive publications, readable by a broad spectrum of the public, that would throw light on the complexity of problems in the Bulgarian society, the Balkan region and in the world.

Acknowledgements

The methodological scheme for executing the content analysis was prepared by researcher Krasimira Trendafilova in collaboration with the author of this chapter. The methodological instructions were carried out by Stela Ivanova under the author's guidance. Natasha Mileva helped with the graphics. All are from the Institute of Sociology of the Bulgarian Academy of Sciences, and their contribution is gratefully acknowledged.

Chapter 20

The Struggle for Djerman-Skakavitsa: Bulgaria's First Post-1989 "Water War"

Caedmon Staddon

In February 1995 the Bulgarian government sent Interior Ministry troops into the picturesque town of Sapareva Banya on the northern slopes of the Rila Mountains to quell a popular protest against a water diversion project designed to help alleviate a water shortage in the capital (see Figure 18.1). Sparked by the Bulgarian state's proposal in late 1994 to resurrect an older water diversion scheme in the Rila Mountains south of Sofia (Voden Kompleks "Rila" or VK-Rila), this protest pitted residents of the water-scarce capital against North Rila communities fearful of potential environmental damage and angered by the lack of government consultation and dialogue. After several rounds of inconclusive negotiations, road blockades and public rallies, the impasse was "settled" by the deployment of Interior Ministry troops and the declaration of martial law in the area on 8 February 1995. The proposed diversion was completed later that year, diverting water from the Skakavitsa River into the Iskar River (see also Hristov *et al.*, this volume).

The conflict over water resources that took place in the Rila Mountains of southwestern Bulgaria in late 1994 and early 1995 raises some important political challenges for a country facing mounting water crises. It also raises philosophical and theoretical questions for scholars attempting to understand environmental protest and conflict in post-communist countries. This chapter will emphatically not follow the line of the government of the day's interpretation of the water protests as "anti-democratic" and "anti-Bulgarian." Rather I argue that these events can be interpreted in terms of the ways in which an ostensibly minor environmental issue, a local protest against a relatively small water diversion project, was constituted as something much larger through its complex interconnections with the *economic, political* and *socio-cultural* dimensions of the current transition from state socialism to capitalist democracy.[1] Put another way, Djerman-Skakavitsa would probably not have

---

[1] Following Swyngedouw (1999: 447) I argue waters of the Djerman-Skakavitsa catchments were constituted as important "socio-natural" actors on a national political stage.

become such a cause célèbre (and subsequently an object of popular memory and analysis) had it not aligned with an economics of "transformational recession", politics of continued central control (under the guise of "democracy") and a socio-cultural experience of exploitation and separateness in the North Rila region. These factors contributed to the constitution of the Djerman-Skakavitsa protests, including within them the roles of "citizens", the state apparatus, civil society and indeed "water" itself.[2] One lesson of this case study is that local-central, state-society, and human-environment relations are dynamics that must be acknowledged in future water management deliberations. Another lesson is that Western scholars must take a more ethnographic approach to the study of change in the post-socialist world (Burawoy and Verdery 1999).

The Economic Background

The atmosphere of general crisis that had pervaded the country since the collapse of the last communist government in November 1989 is reflected in official economic statistics (Table 20.1). Whereas *nominal* GDP increased between 1989 and 1994, *real* GDP, denominated in U. S. dollars, fell by almost 40% as hyperinflation ravaged the domestic economy (a situation that hit the country again in 1997-8–unemployment rocketed from near zero in 1989 to almost 500,000 workers in 1995, although this was somewhat better than the preceding two years). The production sectors were characterized in this period by fairly sharp contraction in levels of output and, more worrying, by accelerated disinvestment of capital resources. This situation was greatly exacerbated by the imminent collapse of the financial sector that was apparent by late 1994 (Angelov *et al.* 1994; Wyzan 1996). Worst of all for general standards of living, consumer price inflation in the mid-1990s was often over 300% (cf. also *BTA Daily Bulletin* January 1997), this in the face of consumer price convergence with Western European levels for many commodities. Unfortunately Bulgaria would experience a second great meltdown, culminating in numerous bank failures, the resignation of the government and further anti-government riots of December 1996-January 1997, and repeated changes of government thereafter.

---

[2]   Elsewhere (Staddon 2000) I have argued for the establishment of a "critical political ecology" that can incorporate greater sensitivity to the constitution of political subjectivities and natural resources as objects of governance within a neoliberal capitalist political economy (cf. Blaikie 2000; Swyngedouw 1999).

Table 20.1 Bulgarian economic indicators, 1989-1996

Indicator	1989	1990	1991	1992	1993	1994	1995	1996
GDP (billion LEVA)	35.6	45.4	131.0	195.0	299.0	543.0	867.7	1700
GDP (billion $US)	17.6	6.9	7.5	8.4	10.8	10.0	13.0	10.8
GDP per capita ($US)	1957	769	836	990	1280	1184	1546	1220
Real GDP (% Change)	-1.9	-9.1	-11.7	-5.7	-2.4	1.4	2.1	-9.0
Unemployment (1000s)	0.0	65.0	419.0	577.0	626.0	488.0	----	478.8
Unemployment rate (%)	0.0	1.6	10.8	15.5	16.4	12.8	10.8	12.5
Average mo. wage LEVA	274	378	1012	2047	3145	4708	7597	12290
Average mo. wage ($US)	136.0	58.0	58.0	88.0	114.0	87.0	109.0	77.8
Gross industrial production (%)	-1.1	-16.8	-22.2	-15.9	-6.9	4.5	----	-6.0
Gross agricultural production (%)	0.8	-6.0	0.0	-12.0	-18.2	0.8	----	----
Gross fixed investment (%)	-0.5	-25.1	-15.6	-26.3	-29.7	n.a.	----	----
Consumer price inflation (%)	10.0	72.5	338.9	79.6	64.0	121.9	132.9	410.8
Producer price inflation (%)	n.a.	n.a.	284.0	24.9	15.3	91.5	----	257.1
Nominal average wage growth (%)	8.7	38.0	167.7	102.3	53.6	51.4	----	161.0

Source: Wyzan 1996; *Bulgarian Economic Review,* various dates 1996

The uncertain economic transition even reached into the water supply sector, where the antics of domestic and international holding companies' attempts to gain control of the recently denationalized and regionalized utilities made for high drama and good newspaper copy.[3] As in other transitional economies foreign water multinationals expressed interest in gaining majority control of Bulgaria's water and sewerage systems (Ruiters 2000; Page 2000; Staddon 2002). The crown jewel of the regionalized water sector, the Sofia Water Supply and Sewerage Company (Sofia ViK), was the subject of at least two foreign and one domestic buyout attempt notwithstanding the perilous state of its physical infrastructure, leaking over

---

[3] Page (2000) discusses similar attention to the Cameroonian water sector by international supervisory bodies and points to the ways in which they result in the "scaling down" of formerly nationalized water supply institutions in ways that virtually guarantee their eventual privatization.

60% of supply between source and consumer, a factor which greatly contributed to the water shortage crisis in the first place (*Bulgarian Telegraph Agency* 14 June 1995). Saur Internationale had been conducting negotiations with Sofia City Council since 1992 for a joint venture that would have given it a 49-51% shareholding and a 30 year license to provide water and sewerage services in the Sofia capital region. At the same time, the city executive branch, led by then-mayor Alexandar Yanchoulev, was conducting parallel negotiations with another company, French Generale des Eaux, resulting in a political clash within city government. The World Bank made institutional restructuring conducive to eventual privatization (e.g., the break-up of the former national leviathan into regionalized water companies) a condition of its 1994 loan of US$98 million to help repair the water infrastructure (see also Staddon 2002).[4] Similarly the International Monetary Fund has long insisted on the ultimate privatization of all utilities as a key element of its structural adjustment program for Bulgaria (*Bulgarian Telegraph Agency* 28 May 1996; Government of Bulgaria 1998; IMF 1998). Finally, domestic interests, such as the large Kremitkovski Ferrous Metallurgical works east of Sofia, sought to secure their own water supply needs through various contractual and other channels. In the event neither SAUR nor Generale des Eaux won control over Sofia ViK, a plum that went to the UK-US joint venture, International Water Holdings PLC, in 1999 (Staddon 2002).

While the situation in the Bulgarian capital had been difficult in terms of both economic growth and the parlous state of the urban infrastructure, it was undoubtedly even more difficult in the non-urban "provinces." Rural development under central planning had traditionally been orientated towards the industrialization of agriculture (through the state farms) and the extensive exploitation of natural resources (Hristov *et al.* 1972; Staddon 1999). Both sectors were hard hit after 1989 by rapid disinvestment and plant closures (Begg and Pickles 1998). Rural unemployment rates were close to double those prevalent in the urban core, a situation made worse by the disinvestment of industrial capital away from rural areas as combinats retrenched behind larger core production units in urban areas. The communist era's "social industries" policy, which sought to locate some ongoing industrial employment in even the smallest villages, effectively collapsed.

This was the case in Sapareva Banya, the locus of the 1994-95 water protests and subsequent police action, where as many as 25% of North Rilans were

---

[4] Subsequently, Sofiiska Voda was created with a 25-year concession to International Water UU (Sofia), whose parent companies include Bechtel Enterprises Holdings Inc., Edison SpA and United Utilities Plc. The group is financed by private funds and a €31 million loan from the European Bank for Reconstruction and Development.

unemployed, having been let go from the industrial kombinats at Dupnitsa, Pernik and Radomir, the soft coal operations at Bobov Dol, or other smaller local industrial plants as successive waves of plant closure overwhelmed the state sector after 1989 (Staddon 1996, 1999). Virtually all state-owned manufacturing in the area had largely ceased operation. Figures for average household income reflect this situation, with average income in Sapareva Banya at about US$750/year, well below regional and national averages (Table 20.2; National Statistical Institute 1993). By 1997 a few of the vacated production facilities had been "recolonized" by smaller-scale producers of textiles, although they provide relatively little reliable employment (Staddon 1999). For many it is only the relatively widespread access to restituted agricultural lands and local common property resources (especially forest resources) that makes the satisfaction of household needs possible, though this land is generally farmed at a relatively low level of productivity, and in the face of numerous difficulties (Dobreva and Kouzoundra 1994; Staddon 1996).

Numerous authors have pointed out how state-socialist development policies saw the environment as an *obstacle* that must necessarily be *overcome* on the path to socio-economic development (Komarov 1984; Peterson 1991; Pickles *et al.* 1993). Development ideology based on "productionism" unambiguously characterized the natural environment as an opponent of, or barrier to, social and economic development: "we must not wait for favours from nature, our task is to take them from her" (quoted in Pickles *et al.* 1993: 170). In the Bulgarian instance it was argued that the state-socialist development model would result in the dialectical reconstitution of *man* as

Table 20.2 Social statistics, Kyustendil Okrug (District), 1992

Municipality	POPULATION			INCOME AND EMPLOYMENT			
				Average Annual Income			Percent state sector
	Total	Urban %	% >65	Leva	% BG average	% Region average	
Boboshevo	4620	39.9	44.9	12495	53.2	56.7	29.8
Bobovdol	13655	56.1	26.7	38200	162.7	173.4	75.8
Dupnitsa	55737	74.3	22.5	27499	117.1	124.8	30.6
Kocherinovo	7500	40.5	38.7	15902	67.7	72.2	29.6
Kyustendil	78328	69.5	24.6	20343	86.6	92.3	32.2
Nevestino	5894	0.0	62.0	18078	77	82.1	20.0
Rila	4410	77.6	33.1	20644	87.9	93.7	22.3
Sapareva Banya	9544	48.6	29.0	18690	79.6	84.8	11.4
Trekliano	1659	0.0	64.4	16929	72.1	76.8	16.8
Sofia Region	986253	59.7	23.7	22032	n.a.	100.0	30.1

Source: National Statistical Institute 1993

well as *nature* in which the latter becomes a perfectly objectified and managed backdrop for the society of "new model men" (Bozhilov *et al.* 1981). Natural resources, such as rivers and forests, were easily reconstituted through this "environmental imaginary" as malleable objects of the will of an preeminent communist society. Thus it was popularly accepted that large dam projects (for water supply and for power) were self-evidently justified and so, to even a greater extent than in Western capitalist countries, the hydro-engineering paradigm was allowed free reign (*cf.* Donahue and Johnston 1998). More than this, large-scale hydroengineering became a high moral imperative and metaphor for the industrial modernization. During the communist era the powerful water development sector built well over 1000 dams, reservoir and diversion projects within the country. Bulgarian hydro-engineers also exported their fervor for "taming water to man's needs" all over the developing world throughout the Cold War period.

The November 1989 anti-communist revolution has not necessarily meant a wholesale change in attitudes towards the natural environment. While opposition to large environmental projects such as VK-Rila were part of the opposition to communist rule in Bulgaria in the 1980s, the mindset that produced them, the "hydro-engineering paradigm," has remained near the heart of environmental decision-making. Though restructured during the 1990s, the hydro-baronies of Vodproekt, Hidrostroi, Energoproekt, the publically-owned Water Supply and Sewerage Companies (ViKs) and the National Electricity Company all remained in place in throughout the 1990s (Kuyumdjiev 1994). By the end of the 1990s these agencies were again proposing large-scale hydro-engineering projects, such as VK-Rila and the "Gorna Arda" in southwestern and southcentral Bulgaria respectively, but this time arguments about modernization were articulated with even more compelling neoliberal rhetoric, using claims about foreign exchange and the "businesses" of water supply and hydroelectricity generation.

Elsewhere I have elaborated on the political economic marginalization of rural areas like North Rila, as manifested in rural disinvestment and in the stripping away of natural and human resources (Staddon 1996, 1999). Here I will concentrate on the implications for the economic aspects of Djerman-Skakavitsa. Two things seem pertinent here. First, there is a certain sense in which Bulgaria is literally and figuratively "consuming itself" in order to satisfy the insatiable pangs of unleashed neoliberal accumulationism. Part of this "autoconsumption" process involves the reconstitution of natural resources, the built environment and even labor power, as objects of justified liquidation. The environment, considered as a ubiquitous input to production under central planning, has been subtly reformulated as a necessary casualty of the imperative to consume and to accumulate. Of course, many Bulgarians are deeply concerned about deteriorating environmental quality. However, the rush to satisfy consumer demands, in the absence of an atmosphere of generalized social trust, creates a situation where cut-throat primitive accumulation wins out over prudent economic or environmental management. It is surprising to find that at the very same moment that Sofia was experiencing a severe water shortage crisis, there was an intense battle for corporate control over the municipal water supply company (Staddon 1996).

Water Crisis in the Capital and Protest in the Periphery

The years 1993 and 1994 had seen record drought in Bulgaria. The Iskar Reservoir serving Sofia, a city of 1.2 million people, was nearly entirely depleted by late 1994 (see Hristov *et al.* this volume). On 21 November 1994, the non-party caretaker government of Reneta Indjova (erstwhile Chair of the Privatization Commission) imposed the first 2:1 (two days off for every day on) water rationing on residents of some areas of the capital (*Demokratsiya* 23 January 1995). Between that time and June 1995 when they were finally lifted, the water regimes would be intensified (to 3:1) and expanded to include most districts within the city. Panicked members of the National Parliament had even suggested that a large proportion of the population should be evacuated from the capital during the summer of 1995 if water supplies did not improve (*BTA Daily Bulletin* 16 February 1995). Despite the initial shock of the regimes, however, the shortage of potable water was neither sudden nor unanticipated. Experts in the water sector, the Bulgarian Academy of Sciences and elsewhere had been warning of imminent shortages of potable water in the capital for more than a decade (*Vecherni Novini* 13 November 1990; *Zemedelsko Zname* October 7, 1986). Indeed, not long after the enormous Iskar Reservoir (also known as "Lake Sofia") was completed in the late 1950s, government water engineers proposed a large system of diversions, reservoirs, channels and pumping stations to tap most of the water from the upper Rila Mountains for Sofia's rapidly expanding domestic and industrial needs (Knight *et al.* 1995). The scale of the proposed project, which called for the diversion of surface waters from the Vitosha, Rila and western Rhodope massifs, is remarkable in European terms, encompassing an area of more than 5000 square kilometres immediately south of the Sofia Basin. By 1989 this plan had already been partially constructed, with the Djerman-Skakavitsa Diversion in the North Rila Mountains earmarked as the next phase in its realization.

Effects of the water regime in Sofia were immediately apparent. Even those living in districts spared the water regime itself were compelled to procure spring water for drinking purposes, as the level of the Iskar Reservoir had sunk well below the level for providing even "conditionally clean" water.[5] Everywhere in the city one saw citizens carrying water bottles, and even large tanks for filling at one of several outdoor mineral spring taps located throughout the city. During the 18 hours of water supply each three days (3:1 rationing) bathtubs, jugs, soda bottles, and even improvised cisterns were filled to cushion the subsequent periods of water shut-off, leading some experts to doubt that the regimes resulted in any real conservation effect at all (*Pari* 8 June 1994). These paradoxes of shortage were compounded by the fact that several regional industrial plants, including apparently the Kremitkovski Metallurgical Works northeast of the city, continued to use potable water throughout the crisis (*Standart* 8 January 1995).

The North Rila residents were challenging re-institution of the VK-Rila diversion scheme in reponse to the shortage that during the winter of 1994-95 with petitions, court action, and blockades. The protesters against the Djerman-Skakavitsa Diversion demanded that the project, a holdover from the communist era, be scrapped, that an environmental impact assessment (EIA) be conducted in accordance with the 1992 *Environmental Protection Act*, and that their own long-standing local water shortages be addressed. The protest movement was in fact comprised of a delicately balanced coalition of the locally rooted Citizen's Initiative Committee for the Protection of Rila Waters (CIC) and more professionally orientated environmental NGOs based in the capital.[6] Popular protest against VK-Rila was further inflamed by the fact that popular agitation against precisely this hydro-engineering proposal had been an important element of anti-communist activities in the autumn of 1989, leading to the collapse of the Zhivkov government (Fotev, this volume; Koritarov 1996; Crampton 1997; Staddon 1996). Blockades erected on 21 December 1994 at two different places on the road running from the town of Sapareva Banya to the site of the proposed diversion received substantial support from other nearby localities, from CIC, and also from national environmental organizations such as Green Patrols and Ecoglasnost.

---

5   See Bardarska and Dobrev, this volume.

6   Elsewhere I have examined this coalition in terms of its inherent instabilities with respect to the conceptualization of the object of protest ("water") and the power dynamics of organizing and conducting protest actions against the state (Staddon 1996, 1998), and in terms of its implications for the emergence of civil society in post-communist Bulgaria (Cellarius and Staddon 2002).

In addition to opposing VK-Rila, local activists in North Rila perceived of their protest actions as an attempt to assert a fuller sense of local autonomy in the face of apparent recentralization of allocative and authoritative power by the central state. Indeed local leaders referred to the protest actions and the CIC as "the only democratic experiment in Bulgaria right now" (Staddon 1996). Considerable pains were also taken by local activists to show that the protests were *not* directed against any particular political group, but rather were aimed at an emerging "democratic centrism" in which all political parties were said to be participating. Thus, the blockades were deliberately established *after* the 18 December 1994 parliamentary elections, in which the Bulgarian Socialist Party (BSP) won an absolute majority, in order to minimize the risk of being interpreted in party-political terms. Similarly, local activists repeatedly emphasised the broad base of support that their activities received from people of all political persuasions. A particular point of contention, the abrogation of environmental impact assessment (EIA) requirements codifed in the 1991 *Environmental Protection Act* first by the non-party Indjova interim government and later by the BSP Videnov government, was widely interpreted as a turn *away* from post-Zhivkov attempts to establish a system of governance based on Western democratic principles. Even the community consultations engaged by national politicians in an effort to negotiate an end to the blockades were structured in such a way as to reinforce these perceptions. The CIC and its supporters were able to articulate their opposition at national and international levels even in the face of deliberate attempts by central government (and not always a Socialist Party government) to stack the political deck against them.

Similarly local activitists repeatedly emphasized the broad base of support that their activities received from people of all political persuasions and the orderly rational nature of their objections, as articulated especially by the CIC. The CIC was formed by citizens of Sapareva Banya and nearby Saparevo on 22 December 1994, shortly after the first barricades had been established spontaneously by some local citizens in an unorganized attempt to block movement of construction equipment into the Rila Mountains (Staddon 1996). In registration documents filed with the district court in Kyustendil there were only 31 official members of the CIC, though meetings regularly drew scores of locals, and the CIC was popularly recognized as the official leader and representative of the protest movement.[7] Participation in the group

---

[7] While there is disagreement about the proportion of the population that supported the positions of the CIC against the Djerman-Skakavitsa diversion, the smallest estimate I received, from the Secretary to the Obshtina Council (Staddon 1996) in interviews was about 60%. He further claimed that "about 40% might have agreed to the diversion, but only if

transcended age, economic and party-political divisions as well as drawing participants from other nearby villages. Leaders of the CIC tended to occupy important social positions within the local community; including the Principal of the Middle School, the coordinator of the local community drop-in centre and a local publican and member of the environmental group Green Patrols. Part local institution (which has continued to organize since the protests in 1995) and part social movement, the CIC depended for its legitimacy on its strong local roots (Dainov 1997). Working through local institutions (middle school teachers organized lessons on surface water ecology during the protests) and local networks (information was often disseminated through the community center, the bars and even the church) the CIC always retained a flexibility and even an ambiguity that was crucial to its success as an organ of local protest. Ironically, it was precisely the prominence and political legitimacy generated by *losing* its initial struggle with the state that has helped underpin its subsequent activism as a social and political institution.

The transition state was unable to recognize that the CIC offered far more than merely a vehicle for organizing opposition to a single state policy. Indeed it is difficult to understand the actions of central government representatives, from both "red" and "blue" political camps, in that period without the assumption that they saw the situation exclusively in terms of parochial concerns of the residents of North Rila and the instrumental rationality of the Djerman-Skakavitsa scheme. Had they seen the CIC more in terms of deeper (and more enduring) political and civil reconfigurations and principles, it is possible that they might have behaved differently. In lieu of broader understanding, the consultations that government politicians offered the CIC became very much a reflection of the urban-oriented hegemony that was reasserting itself in the economic sphere. Several consultations were held between representatives of national government and local diversion opponents between the end of November 1994 and the beginning of February 1995. Interestingly, as the meetings dragged on into January and February 1995 government representatives sought to disarticulate the consultations process from the local community, and from the CIC in particular. After a move from the mayoralty building in Sapareva Banya to the nearby mountain resort of Panichishte, a further attempt was made to restrict participation by declaring that only members of the Bulgarian Socialist Party could attend. Further meetings were proposed for BSP headquarters in Sofia, while legal challenges by the CIC took the deliberations into the regional courts at Kyustendil, 50km distant. Countering this dislocation of the places

---

their water problems were solved also." Most other informants claimed that support for the CIC was much higher.

of formal dispute, the leadership of the CIC organized public rallies in Sapareva Banya and Dupnitsa in order to keep the situation in the public eye, to keep local people's minds focussed and to expand the geographical bounds of support. Ultimately these spatial dislocations increasingly came to mirror shifts on both sides towards more extreme positions.[8]

Key representatives of the government side included interim Prime Minister (October 1994-December 1994) Reneta Indjova, BSP Prime Minister Zhan Videnov and his Regional Development Minister Doncho Konakchiev (who succeeded Indjova after the 18 December 1994 national elections). Also vocal, though not a direct participant in the formal consultations, was Sofia Mayor Aleksandar Yanchoulev, who repeatedly called for central government to ensure the rapid construction of the Djerman-Skakavitsa diversion.[9] These core orientations and urbanization effects were also quite sharp *visually*. Documentary footage assembled by the environmental group Green Patrols and Bulgarian National Television shows the shiny black Mercedes of the carefully coifed Ministers, Deputy Ministers and their entourages pulling up to designated meeting places amid the horse carts, Wartburgs and Ladas of the townspeople. Once in the town centre these urbane representatives of Sofia's interests were faced with the scores of locals; men in tattered work clothes and women in old print dresses and scarves who supported the CIC. All of these factors assisted to impose of the will of central government and to marginalize the CIC. Perhaps it is therefore not so surprising that the dialogues themselves were also remarkable for the complete failure of the two sides to engage one another. Certainly there seems to have been very little common ground upon which to establish a collective understanding of the water crisis that did not ultimately appeal to the state's police monopoly.

On 8 February 1995 special troops from the Bulgarian Ministry of the Interior were deployed to break through the two-month-old barricade that local citizens had put across an access road into the Rila Mountains. This was deemed necessary in order to guarantee the access and security of construction crews who were to construct the pipeline diverting water from the Skakavitsa River to the Djerman River, a distance of some 3 kilometres, and thence into the Sofia water system. Squeezed between demands from the capital for more water and from the "provinces" for more democratic process, the central state opted decisively to satisfy the capital. In an action

---

[8]   There were those who talked darkly of resorting to direct action tactics, especially after the central state showed willingness to resort to military force on February 8.

[9]   Yanchoulev was in fact embroiled in a scandal over the privatization of the Sofia Municipal Water Supply and Sewerage Company and his handling of the water shortage at the time (Staddon 1996).

subsequently referred to by the Interior Minister as a "textbook model of police work guided by principles of democracy and respect for human rights" nineteen people were arrested and about a dozen were injured, including one person seriously. Several journalists had their cameras confiscated and their film destroyed by police (Staddon 1996). A large police force remained in the area through the end of March 1995, guarding the road and harassing locals (*BTA Daily Bulletins* 13 February 1995, 24 March 1995; Staddon 1996).

Much more needs to be said about the relations between the CIC and the central state before the analysis can be understood to be in any sense "complete." The barricades themselves explicitly created an oppositional space constituted through discourses of motherhood, nation and soil. Flags were flown, popular nationalist folk songs were sung and the front-line was often made up of "Babas," Bulgarian grannies. The somewhat strange convergence of the politics of the BSP and the opposition Union of Democratic Forces (UDF) over Djerman-Skakavitsa (with of course a few notable exceptions) was also significant, bespeaking more than merely a tacit agreement about ends. Interestingly it appears that the mode of governmentality itself, of the practice of government and of the state's rightful relation to the population and objects (such as water resources) it administers, actually appears to have *united* both sides of the basic BSP/UDF split against this expression of local political autonomy. The national UDF refused to officially condemn the military intervention at Sapareva Banya, even while it simultaneously interpreted Djerman-Skakavitsa as an attempt to curry the favor of voters in the capital (*Demokratsiya* 19 February 1995) Clearly, the meetings between government and local representatives and the protests themselves constituted strategic interactions of different modes of spatial contestation: core versus periphery, urban vs. rural.

The Cultural Politics of Protest Against Djerman-Skakavitsa

As noted earlier in this chapter, there has been a revalorization of the *urban* in Bulgarian social life, not merely as the locus of industrial production (as was the case under central planning), but also as the locus of the objects of post-communist desire: flashy cars, clothes, music, lifestyle, fast food, etc. The withdrawal of industry from the countryside has meant, among other things, that relative economic prosperity is only encountered in the cities, and especially in the capital, which therefore becomes a magnet both for the rural unemployed and for the consumer goods that have become a national fetish (it is even possible to purchase replica designer labels to stitch on the outside of one's non-designer clothes in some open air markets in the city). Thus the (re)configuration of identity through material consumption as a key element in post-communist subjectification has implied *a revaluation of urban spaces*, and especially capital cities, as the most important spaces of transition capitalism (Buck-Morss 1989; Staddon and Mollov 2000). Largely in concert and simultaneous with the political and economic reconfigurations, strategically significant cultural symbols were used to bolster each side of the Djerman-Skakavitsa issue. Not just the objective realities of being "peripheral," or "rural" but also their relative valuations in Bulgarian popular culture were interpreted and deployed strategically in complex and contradictory ways.

The simultaneous activation *and* denigration of rural symbologies is a long-standing paradox in Bulgarian popular culture. Bulgarian national consciousness is closely bound up with images of rural life, agricultural production and what the martyred Bulgarian leader Aleksandar Stamboliiski castigated as the "pernicious artificialities of urbanism" (Bell 1977).[10] Notwithstanding communist-era ideology about the "new model man" as an ideal *urban industrial* worker, Bulgarian literature frequently invokes the image of the crafty rural peasant farmer (e.g., "Bai Ganyo" and "Hitar Petar") as the backbone of the Bulgarian nation (*cf.* Sanders 1949; Todorova 1997). The national flag itself, unfurled so proudly at the barricades, also celebrates Bulgaria's rural-agricultural origins, the green swath denoting "the fertility of the Bulgarian land and the love of the Bulgarian people for their homeland" (Bokov 1981:15). Although the rural-to-urban food transfers again became critical during the 1990s, forming a large "second economy" through which Bulgarians impoverished by the economic transition sought to fill domestic consumption (Creed 1993; Dobreva and Kouzoundra 1994), there is a reflexive devaluation of rural identities and lifestyles, expressed in the popular aphorism "Better to be second in Sofia than first in my village" (Creed 1993). Thus, in the case of Djerman-Skakavitsa rural identity played a specific role on both sides of the issue; identification of opposition as "rural" in urban discourse was generally pejorative while equations between opposition and rurality were celebrated locally as proof of the authenticity of the CIC action.

While it was often claimed that the opposition of North Rilans proceeded from an untutored "peasant" mentality, there was a need to appeal to other means of political division. Because most Bulgarians value their connections with rural localities highly (both materially and symbolically) it

---

[10] Stamboliiski led the populist and democratically elected government of the Bulgarian Agrarian National Union (BANU) between 1919 and 1923. The 1923 military coup that deposed his government and executed him is popularly believed to have marked the end of democratic government until 1989.

was necessary to radically reinterpret the specific rural identity in North Rila. Very important in this regard was the symbolic coding, "rural=communist" and "urban=anti-communist," which had become a basic element of boilerplate social and political analysis in Bulgaria, invoked to "explain" everything from BSP electoral strength, to the hardships faced by city-dwellers specifically and to the strains of transition generally. Thus, local opposition to Djerman-Skakavitsa was sometimes attributed by urban analysts and central government representatives to the machinations of the pro-communist "provincials" and their crypto-communist leaders (Staddon 1996; *Demokratsiya* 10 February 1995).[11] Creed (1993) further notes an unfortunate tendency for popular urban discourse to conflate the rural with the anti-urban, and even the anti-Bulgarian. Interim Prime Minister Indjova clearly invoked such stereotypes by declaring that protesters against Djerman-Skakavitsa must be "separatists" who didn't care if "Sofia died of thirst" (Staddon 1996), though many other examples can be cited. Survey research reported by Fotev (this volume) suggests that this discursive coding did not entirely work in the country as a whole: popular support for the Djerman-Skakavitsa diversion was never more than about 27% of the adult population. Not surprisingly representatives of the BSP government elected in December 1994 were forced to alter these symbolic associations, arguing that it was the anti-BSP "blues" who created this situation by strongly identifying Bulgaria-in-transition with urban areas and with urban cosmopolitanism over and against rural political associations (*Bulgarian Telegraph Agency* March 1995).

In contrast, the members of the Civil Initiative Committee were attempting to articulate a marginalized rural identity with a nationalist/ized sense of the possibilities for Bulgarian democratization. Deployment of nationalist and rural symbologies at the barricades was therefore powerful, attempting to unite a set of local actions in a specific place with an enduring sense of Bulgarian national identity. Many protesters at the barricades sang popular Bulgarian folk songs as the troops massed against them. Time and again in the series of governmental consultation meetings called to discuss Djerman-Skakavitsa, local protesters castigated national politicians for being "too far from the village" in their policy making process (Green Patrols 1995). And locals tended to talk about their actions in terms of popular characteristics of national identity such as stoicism, forbearance and will and the myriad cultural/religious practices associated with water (Fotev, this

---

11 Linguistically, the terms for "provincial", "provintsialen" and "province", "provintsiayata" are commonly used to mean any area outside the urban core of the country, and any attitudes or ways of life that differ from the supposed cosmopolitanism of Sofia. The pejorative content of the term is often uppermost in its connotations.

volume). Thus the central state negativity about rural identity was inverted to become positive, along the way picking up certain meanings and symbolic associations from local historical-geography, environmentalism, and democratization. But in full concurrence with Creed (1993) it must be noted that the murkiness of the situation is exacerbated by the wide gulf of cultural misunderstandings and even propaganda that separate the anti-rural views of the urban core and the anti-urban views of the rural periphery.[12]

Response to Environmental Crisis

The Djerman-Skakavitsa struggle has lessons for both the theorists of political transitions in post-socialist states and for Bulgarian environmental managers. It is tempting to conclude that the police crackdown on the Djerman-Skakavitsa protests constituted a significant failure of the post-communist democratization process. The citizens of North Rila failed spectacularly in their attempts to force the state to change its plans to construct the Djerman-Skakavitsa Diversion, even to get the state to follow its own legally mandated requirements for EIA and real community consultations, let alone promised local water improvements. The state (in both "red" and "blue" guises), for its part, was unable to reconcile its all too evident will-to-power with norms of democratic negotiation and accommodation. However, a more fruitful way of framing the issues is in terms of the changing nature of power's relation to its objects of application and administration. The case of local protest discussed in this chapter is interesting precisely because it exposes the ragged juncture between the dominant state and local autonomy and resistance. At the very least the people of North Rila proved again that it is possible to resist the pressures of the state and to achieve some limited gains even if the bigger issues are (temporarily) lost. Local citizens were able to organize the CIC, which has since 1995 evolved into the Foundation-Sapareva Banya with a primary mission orientated towards local development. Local political mobilization also proved sufficiently robust to replace the pro-diversion mayor and some of the Municipal Council in the October 1995 local elections. To some extent, therefore, limited political gains have been institutionalized. More far-reaching however is the fact that a local political identity was articulated with the experienced realities of economic, political and socio-

---

[12] There was also an ethnic dimension which is not elaborated here—in which Sofia residents viewed people from North Rila (so-called Pirin Macedonians) as "naturally fractious" and where the same people used their Macedonian heritage as evidence of successful opposition to unacceptable central authority (Staddon 1996).

cultural marginalization that provides a positive model for residents of the North Rila Region. As Pred and Watts (1992:18) point out: "it is the *production of cultural difference within a structured system of global political economy* that is central to an understanding of both the territory-identity relation and to the reworking of modernity." Case studies such as Djerman-Skakavitsa show us how certain discourses about "democracy" can, somewhat ironically, actually close down the spaces for legitimate challenges to power's exercize in more complete and subtle ways than the authoritarian politics that preceded them (cf. Foucault 1991). Even so, it remains possible for local people to articulate their challenges, here through direct identification with landscape and reaching out to political discourse at other scales, national and international.[13]

From the viewpoint of environmental management and coping with future drought in Bulgaria, this chapter focuses on the periphery and conflicts with central authority. In coping with future drought—or with climate change for that matter—what will be the respective roles of central versus regional, local, or river basin authority? How will those roles differ in response to emergency versus longer-term environmental planning? What will constitute "due process" in dealing with disputes that range across geographic, social, and political space and scale? Will crisis remain an excuse to resurrect a paradigm of resource management without examination of longer term issues and consequences? Worrisome surely is the post-communist state's apparent inability to address water concerns in anything other than an "emergency" mode – especially since 2000 and 2001 saw renewed drought and the potential for rationing regimes in Gabrovo, Blagoevgrad, Vratsa, Montana, Stara Zagora and other provincial towns. Finally, what will be the locus and nature of decisions that must balance (or give priority) among environmental quality, economic development, and emergency preparedness? The "deep blue" government of Ivan Kostov announced its intention to go ahead with the highly controversial Gorna Arda hydro-electric scheme in the Rhodope Mountains.[14] These are questions that must be answered by Bulgarians in the context of the transition they are experiencing. As the Djerman-Skakavitsa "water war" illustrates, the economic, social and political processes that will answer these questions are in flux. Recognizing the relevant challenges may point to some directions of decision making that Bulgarians can view as fair, equitable, and open.

---

[13] This latter is of course one of the time honored "weapons of the weak" – articulating their problems with the concerns of distant others.

[14] The government of Simeon Saxe-Coburg-Gotha, which replaced Kostov's UDF government in June 2001, has not, so far as I am aware, repudiated Gorna Arda.

References

Angelov, I. *et al.* 1994. *Economic Outlook of Bulgaria 1995-1997.* Sofia: Institute for Economics Bulgarian Academy of Sciences.

Bauman, Z. 1992. *Intimations of Postmodernity.* New York: Routledge.

Begg, R. and J. Pickles. 1998. "Institutions, Social Networks and Ethnicity in the Cultures of Transition: Industrial Change, Mass Unemployment and Regional Transformation in Bulgaria", pp. 115-146 in J. Pickles and A Smith (editors), *Theorising Transition: The Political Economy of Change in Central and Eastern Europe.* London: Routledge.

Bell, D. 1977. *Peasants in Power: Alexander Stamboliiski and the Bulgarian Agrarian National Union 1888 - 1923.* Princeton, NJ: Princeton University Press.

Blaikie, P. 1999. "A Review of Political Ecology: Issues, Epistemology and Analytical Narratives", *Zeitschrift für Wirtschaftsgeographie*, 43(3-4):131-147.

Bokov, G. (editor). 1981. *Bulgaria: History, Policy, Economy, Culture.* Sofia: Sofia Press.

Bozhilov, I. *et al.* 1994. История на България. София: Христо Ботев.

Buck-Morss, S. 1989. *The Dialectics of Seeing: Walter Benjamin and the Arcades Project.* : Boston, MA: MIT Press.

Bulgarian Telegraph Agency. Дневен бюлетин (*Daily Bulletin*), various dates, Sofia.

Burawoy, M. and K. Verdery (editors). 1999. *Uncertain Transition: Ethnographies of Change in the Postsocialist World,* Rowman and Littlefield.

Crampton, R. J. 1997. *A Concise History of Modern Bulgaria.* Cambridge: Cambridge University Press.

Creed, G. 1993. "Rural-Urban Oppositions in the Bulgarian Political Transition," *Sudosteuropa* 42:369-382.

Dainov, E. 1997. Власти и хората, доклад на конференция "ВК Рила и гражданското общество," Сапарева баня, м. юли 1997 (unpublished mimeo).

Demokratsiya (Демокрация, Bulgarian daily newspaper). 1995. Sofia (various dates).

Dobreva, S. and V. Kouzoundra. 1994. "Some Problems in the Transformation in Bulgarian Agriculture," pp. 77-102 in N. Genov (editor), *Sociology of a Society in Transition.* Sofia: Regional and Global Development Institute, Bulgarian Academy of Sciences.

Donahue, J. and B. R. Johnston. 1998. *Water, Culture and Power: Local Struggles in a Global Context.* Washington, DC: Island Press.

Foucault, M. 1991 "Governmentality," pp. 87-104 in G. Burchell *et al.* (editors), *The Foucault Effect: Studies in Governmentality.* Chicago: University of Chicago Press.

Government of Bulgaria. 1998. *National Strategy for Accession of the Republic of Bulgaria to the European Union,* Sofia, March 1998

Government of Bulgaria. 1991. *Environmental Protection Act,* Sofia: Government Printing House.

Green Patrols. 1995. Документиране на протестите за "Джерман-Скакавица." София.

Hristov, T. *et al.* 1972. Икономическо-географски характеристики на Кюстендилски окръг относно структурното развитие на микро ниво. Годишник на Геолого-географския факултет при Софийския университет, София, стр. 64.

IMF. 1998. Letter of Intent of the Government of Bulgaria and associated IMF policy documents, published on the IMF website.

Knight, C. G., S. Velev and M. P. Staneva. 1995. "The Emerging Water Crisis in Bulgaria," *GeoJournal* 35(4):415-423.

Koritarov, V. 1996. Рилски разкази от долините на Джерман и Скакавица. София: Илинда-Свтимов.

Kuyumdjiev, N. 1994. Международна рамка на ВиК сектора в България. София: МОСВ.

Ministry of Environment. 1991. Закон за опазване на околната среда. София.

National Statistical Institute. 1993. Областите и общините в Република България. София.

Page, B. 2000. "Water, Citizenship and Community Development in Cameroon", presented in "The Politics and Policies of Natural Resources in Developing Areas," special paper session held at the Annual Meetings of the Royal Geographical Society/Institute of British Geographers Annual Meetings, University of Sussex, Brighton, England.

Pari (Пари, Bulgarian daily newspaper), 8 June 1994.

Peterson, D. J. 1993. *Troubled Lands: Environmental Pollution in the Former Soviet Union*, Santa Monica, CA: RAND Corporation.

Pickles, J. and the Bourgas Group. 1993. "Environmental Politics, Democracy and Economic Restructuring in Bulgaria," pp. 167-185 in J O'Loughlin and H van der Wusten (editors), *The New Political Geography of Eastern Europe*, London: Bellhaven Press.

Pred, A. and M. Watts. 1992. *Reworking Modernity: Capitalisms and Symbolic Discontent*. New Brunswick: Rutgers University Press.

Ruiters, G. 2000. "TNCS and Water Privatisation in South Africa," presented in "The Politics and Policies of Natural Resources in Developing Areas," special paper session held at the Annual Meetings of the Royal Geographical Society/Institute of British Geographers Annual Meetings, University of Sussex, Brighton, England.

Sanders, I. 1949. *Balkan Village*. Lexington: University of Kentucky Press.

Staddon, C. 1996. *Democratisation, Environmental Management and the Production of the New Political Geographies in Bulgaria: A Case Study of the 1994-95 Sofia Water Crisis.* unpublished Ph.D. dissertation, Department of Geography, University of Kentucky.

Staddon, C. 1998. "Democratisation and the Politics of Water in Bulgaria: Local Protest and the 1994-95 Sofia Water Crisis," pp. 347-372 in J. Pickles and A Smith (editors), *Theorising Transition: The Political Economy of Change in Central and Eastern Europe*. London: Routledge.

Staddon, C. 1999. "Economic Marginalisation and Natural Resource Management in Eastern Europe," *The Geographical Journal*, 165(2):200-208 (also in Bulgarian, in *Problemite v Geografia*, 1999, 1).

Staddon, C. 2002. "Privatisation and Access to Water Resources in Postcommunist Eastern Europe" at *Riversymposium 2002: The Scarcity of Water*, 3-6 September 2002, Brisbane, Australia.

Staddon, C. and B. Mollov. 2000. "City Profile: Sofia, Bulgaria," *Cities: The International Journal of Urban Policy and Planning*, 17(3):379-387.

Standart (Стандарт, Bulgarian daily newspaper), 8 January 1995.

Swyngedouw, E. 1999. "Modernity and Hybridity: Nature, Regeneracionismo, and the Production of the Spanish Waterscape, 1890-1930," *Annals of the Association of American Geographers* 89:443-465.

Todorova, M. 1997. *Imagining the Balkans*. New York: Oxford University Press.

Vecherni Novini (Вечерни новини, Bulgarian daily newspaper), 13 November 1990.

Wyzan, M. 1996. "Stabilisation and Anti-Inflationary Policy," pp. 77-106 in I. Zloch-Christy (editor), *Bulgaria in a Time of Change: Economic and Political Dimensions*. Aldershot, England: Avebury Press.

Zemedelsko Zname (Земеделско знаме, Bulgarian daily newspaper), 7 October 1986.

Acknowledgements

I would like to acknowledge the support this research has received from the National Science Foundation (SBR 9515244), the Social Science and Humanities Research Council of Canada and the University of Kentucky Graduate School.

PART IX
LEARNING FROM THE DROUGHT

Chapter 21

Drought in Bulgaria:
What Have We Learned?

C. Gregory Knight, Ivan Raev

In our research we have learned some important lessons about building our analysis of the drought in Bulgaria and in using it as an analogy of future climate change. These lessons are suggestive of the kinds of challenges in analog thinking mentioned earlier. We have not been able to answer all the questions we put to ourselves as well as we would like. It is helpful to us to understand why.

Our reflections of the experience of this work focuses on three levels. First, what did we learn about the drought itself? Here we try to summarize the observations and analysis of the foregoing chapters. Second, if an event such as the drought, and particularly the ensuing crisis of 1994-1995, were to occur again, what would we recommend to the scholarly community in terms of documenting the event and preparing for subsequent analysis? Finally, what have we learned about the strategy of analogy in general?

The Drought

For the past century three continuous drought periods emerged in Bulgaria: 1904-1908, 1945-1953, 1982-1994. The first period was the shortest—20% of the years were dry. During the second period 40% of the years were dry and in the third, over 50%; drought occurred in almost all seasons. The driest years throughout the century were 1907, 1965, 1985 and 1993, with two out of the five driest years being part of the third and strongest drought period. According to existing models of climate change, the coming 50 to 100 years could see a significant rise in temperature, a potential fall in precipitation and almost certain diminished runoff compared to or worse than drought parameters characteristic of the 1982-1994 period. In this case the consequences of this drought could be indicative for the global warming. The experience of this last drought would be exceptionally useful for creating action plans aiming at a reduction of negative consequences.

Throughout the 1982-1994 drought, a 31% fall in Bulgarian runoff was registered compared to the data for the 1890-1996 period. This fall was greater in the runoff basin of the Danube, and less for the Black Sea. The fall in the runoff led to the dangerous decline of water levels in reservoirs and was a major reason for a limited supply of drinking water. Particularly dangerous for drinking water quality were the cases of reaching the "sanitary minimum" limits, a permissible minimum in reservoirs. It was found that in many water treatment stations suitable anti-turbidity, oxidation reagents, reagents for manganese, iron and phytoplankton were not used. No information was given to the public on the real state of drinking water quality nor recommendations for a reduction of risks, such as "boil water" orders.

The drought period demonstrated the importance of setting water use priorities. Had the large reserve of capacities of reservoirs (6 billion m³) and a reasonable regime of water use been maintained, no water crisis would have been reached. If these waters are used for power generation without multiple use during water crisis periods, it is possible that serious problems with public water supplies may occur (for instance, the water shortage in Sofia, 1993-1995). There is a considerable room for improvements of water supplies in Bulgaria, as it has been established that badly maintained water supply facilities had over 50% (sometimes approaching 70%) water losses. Therefore even in periods of insufficient precipitation, the problems of water supplies may be due not so much to water shortages, rather than to improper management.

The 1982-1994 period saw a rise of mean annual temperature by approximately 2°C in the low-lying forest zone. With the increase of altitude in particular in the zone of coniferous woods the warming was from 1.0 to 1.4°C. For the same period a reduction of precipitation from 12.0% to 15.9% was registered in coniferous forests and 24.9% to 28.7% in the zone of oak forests. A survey established no problems arose in water supplies for the period in natural coniferous and beech forests. However in the zone of the oak trees a chronic deficit in humidity was established with consequences for vegetation. The health of natural forests of *Pinus sylvestris, Picea abies, Fagus sylvatica, Pinus peuce,* and *Pinus nigra* did not show substantial negative changes in the drought period. Forests of *Abies alba Mill, Quercus petrraee, Pinus nigra* and others in low lying zones show considerable damage, which is reduced with normalization of precipitation.

In managed forest cultures, especially in the low lying zone of the country (below 800m), and chiefly beyond their natural range, some 162,000ha or about 18.5% of the new coniferous forests in Bulgaria are dying. The reason for these losses is the presence of an unfavorable climatic environment and the weakened potential for coniferous species survival beyond their geographic range. There was an explosion of forest fires losses during the drought period: in 1993 the area of burnt out forests grew by a factor of 28.7. Drought left a serious imprint on agricultural production during the period of study. Four of the six of the driest years out of the last 30 years occurred in the 1982-1994 period: 1985, 1990, 1992 and 1993. In 1993 alone, Bulgaria incurred losses to the amount of US$72.3 million from low yields from wheat and US$102.2 million from the yield of maize. These are followed by losses from yields of tobacco, cotton, barley, sunflower etc. The total direct losses in cereals in Bulgaria in 1993 were US$258.9 million. The study shows that precipitation during the July-August of the current year was of greatest significance for the maize crop, whereas winter grains are most sensitive to winter stores of humidity, and to rain in April and May. Low yields in agriculture have caused higher prices in animal production, making the agricultural sector a poorly competitive sector in general. Considering the fall in wheat and maize yields, affecting the processing industry, live-stock breeding, as well as export, clearly the losses are great.

Changes have also been observed in the behavior of mammal pests whose populations are on the increase in periods of warming. A case in point is the population numbers of voles, which reached record numbers in the winter of 1987-1988. In northeast Bulgaria vole colonies caused considerable damage to the autumn crops at that time. Naturally vole population is also related to other factors, for instance the biological cycle; however warming is a factor contributing to an increase of their numbers.

The drought period of 1982-1994 in Bulgaria had serious economic consequences. There can be no doubt that agriculture and forestry were most affected. If we take the fall in production by 20% owing to drought due to reduced annual timber growth, which is quite realistic, this would mean that the loss from the growth rate of forests alone amounted to approximately 2 million m³ timber annually. If we add on average the 10,000-15,000ha of forest fires annually, it is evident that economic losses in the forestry sector are enormous.

Considerable losses occurred in the hydro-power generation sector, coming from the shorter use of hydro-electric power stations. For the 1980-1992 period power generation from hydro-electric power stations fell from 3713 to 2063 million kWh annually.

Continuous drought resulted in a fall in water quantities in 8 out of the 12 largest reservoirs for drinking water, reaching the critical level, resulting in water rationing in all cities with insufficient water resources. Major cities had water rationing, even in the winter (Varna, Burgas, Pernik, Dupnitsa, Montana, Vratsa, Haskovo, Sofia and elsewhere).

A significant link has been established between the water regime and the rise in incidence of viral hepatitis A and shigellosis dysentery. Shortages of epidemiologically safe drinking water enhance the spread of these diseases, especially when a large part of the population has intermittent water flow.

During the drought period, the theme of water was increasingly raised by the media. Sharp and critical articles appeared, analyzing the causes for the introduction of water rationing and outages. It became clear that Bulgaria had the lowest price of water in Europe, contributing to its wastage. It was reported that over half of the drinking water was lost because of old water supply systems. Independent experts and specialists proved that the water crisis in Sofia was caused by totally impermissible draining of the Iskar dam for the purpose of generation of power, with the covert aim to prove the need for the construction of new water derivations.

The media gave wide coverage of social and political conflicts, provoked by mistaken steps of the Government. An opinion was formulated that the water crisis was essentially due to short-sighted policies and the pursuit of vested interests. Unfortunately the temporary interest towards the consequences of drought quickly died down. The media and the so called "political elite" turned to other social problems. This shows that Bulgarian society still is not aware of the depth of the problem it is facing.

Finally, the drought and ensuing water crisis brought to light the challenges of centralized versus decentralized resource planning and of center/urban versus periphery/rural priorities in resource acquisition. It was, in some dimensions, a challenge to the new democracy in Bulgaria.

Thus, in our study we have learned much about the drought events and about the cascading impacts through environment, economy and society. Bulgaria survived this drought, and the crisis stage finally came to an end with fortuitous weather changes. Could the next major drought be even worse? Will Bulgaria be prepared?

Planning for Future Extreme Events

In our study of the drought, we found that we are both too close and too far from the events to be completely satisfied with our analysis. We are too close in that a scholarly literature and thoughtful consideration of the drought is just beginning to emerge (Knight *et al.* 1995), so we have had to rely on our own work in the first instance. We are too far, in that many things we would have wanted to document or understand in contemporary time are now impossible—how valuable it would have been to undertake social and economic surveys focusing on the drought while it was occurring; how valuable to interview major decision-makers then and subsequently as they reflected on events. Instead, we have relied on a small number of studies focused on the drought and on many of the routine data collection and analytic tasks that with retrospection allowed dimensions of the drought to speak for themselves. Nevertheless, we feel reasonably confident that we have captured the major dimensions of the events, hopefully providing sufficient provocation that other researchers will take up where we leave off, in both spatial and temporal dimension and in depth of understanding.

If an event like the drought was to occur again, we would recommend the immediate creation of a collaboratory of scholars, stakeholders, and policy- and decision-makers to design and execute a plan for monitoring the event and its impacts. Such a group might include people intimately involved in dealing with the drought, but as a body it should be independent of a management or regulatory role. Many of the kinds of data we cite in this book would be the kinds of information the group should gather. As events unfold, other issues will emerge that a plan might not anticipate. In Sofia for example, we have no data about the impact of water rationing on small businesses, although at the time the media spoke of disruptions to bakeries, for example. Anecdotal evidence showed long queues of elderly people at public springs, presumably sent by the extended family to ensure safe drinking water. Similarly, we know people stored large quantities of water in home containers, but how much and to what efficacy is unknown. Also, ridership on public transport may have measurably increased as people made daily pilgrimages to bathe in the homes of friends and relatives not currently on rationing. Some epidemiological data was available to indicate threats to health, but untreated cases of dysentery might have only been indicated by school and workplace absences. To our knowledge, no national surveys of opinions about and actions taken in the face of the crisis were undertaken. Most important, however, is securing data that during the drought were considered confidential, such as the data on water allocations, that continue to fuel speculation on the dimensions of purported mismanagement of Sofia's water.

On the Strategy of Analogy

In terms of the drought as an analogy to climate change, we did identify some important conclusions:

1.   Our analogy has been built on real data and real events. There can be no question that the climate experienced could be a reality: it was;

2. Impacts of the drought were geographically and socially contingent, although the kinds of impacts experienced were not peculiar to Bulgaria. Although Bulgarian society and economy will change markedly by the time that yesterday's drought conditions become a climate norm, many of the underlying historical and cultural precedents will remain the same;

3. This study illustrates the challenges society will face should such events occur again or become more frequent in the future. It also points out large gaps in the social and institutional structure for coping with such events. For example, what might have happened had the drought not broken by mid-1995 and had Sofia's reservoir become completely emptied?

4. Understanding impacts of the drought provides the national public audience with an opportunity to have a glimpse of future climate consequences, thus giving impetus to widespread learning about and participation in planning for climate change;

5. Government, water managers, and citizens appear to have been ill-prepared for the peak drought of 1993-1994. Warnings of impending crisis were evidently suppressed; no transparent discussion of water allocation priorities was ever brought to the public. People were largely left to fend for themselves, with little advice on practical issues of water conservation or health warnings. Only years later was water system leakage addressed, repairs and replacements that could have precluded rationing and the potential of catastrophe. What in Bulgaria is referred to as a water "regime" is a form of rationing by complete interruption of the water supply with concomitant risks to public health; there were no mechanisms to enforce decreased usage with supply continuity. All of these shortcomings must be addressed in the face of climate change;

6. There appears to be no official contingency plan for the emergence of such a drought again, suggesting that there is again risk of *ad hoc* incremental crisis management rather than well-considered steps toward dealing with an emerging issue as it evolves. If there is no drought contingency plan, then responses to the threat of longer term climate change are even more in doubt;

7. Problematic, however, has been our inability to address the drought and potential climate change from the viewpoint of decision-makers. We can certainly detect the tug-of-war between water conservationists and dam- and diversion-builders, and we can point decision-makers to concepts of integrated water resource management to bridge such gaps in the future. The analogy would have been strengthened—and still could be in future planning—by direct involvement of stakeholders and decision-makers, and when it comes to water, that means just about everybody!

8. Whereas water is an important contemporary issue—a new national water plan is underway as this book is being written—climate change lags far behind other exigencies of life in Bulgarian public concerns (O'Connor *et al.* 1999). Perhaps continuing periods of record warmth and repeated drought will change public awareness.

Conclusion

All this said, as a community of scholars we feel that our work points to the importance of planning for drought as a way of adaptation to future climate change. We have no illusion that every year will bring drought as severe as 1993-1994 nor, conversely, that Bulgarian society will have learned to adapt to climate regimes that we label as drought. What we do know is that had the drought occurred in 1989 or earlier, it would have been a certain disaster given Bulgaria's then-prodigious appetite for water (Knight *et al.* 1995). Should Bulgaria redevelop as a nation as dependent on water subsidies to agriculture and industry as it was at that time, it faces almost certain catastrophe with climate change. Drought planning now may lead to a Bulgaria far more resilient to climate change. This we learn by analogy.

References

Knight, C. G., S. Velev and M. P. Staneva. 1995. "The Emerging Water Crisis in Bulgaria," *GeoJournal* 35(4):415-423.

O'Connor, Robert E., Richard J. Bord, Ann Fisher, Marieta Staneva, Veska Kozhouharova-Zhivkova, Stanka Dobreva. 1999. "Determinants of Support for Climate Change Policies in Bulgaria and the USA," *Risk Decision and Policy* 4(1):1-16.

PART X
THE FUTURE

Chapter 22

Recommendations for Policymakers[*]

Ivan Raev, C. Gregory Knight

The consequences of the 1982-1994 drought in Bulgaria are models of possible negative changes we could expect over the next century from the continuing process of climatic change caused by the global warming. Seen from this aspect we have the rare chance to discuss the consequences of both the drought itself and similar phenomena which may be forthcoming. Can Bulgaria create a strategy to minimize the impacts of future drought and climate change?

In this final chapter we try to summarize recommendations that derive from our study, hopefully to assist policy- and decision-makers towards taking useful advanced action. In this sense, this section could be seen as an addition to the *National Climate Change Action Plan* (NCCAP 1999) adopted by the Council of Ministers of the Republic of Bulgaria in 2000. Our study complements the NCCAP by taking an integrated assessment of the drought as an analog for climate change, with obvious focus on water. As we noted in the previous chapter, the work reported in this book results from retrospective analysis by scholars; we have not been driven by the questions or concerns of stakeholders or decision-makers except as we anticipate the concerns of such individuals. Nevertheless, we believe our work can help Bulgaria address an uncertain environmental future.

In our recommendations, we are also motivated by a concern for the low level of public awareness and knowledge of global warming and climate change impacts in Bulgaria. In a study documenting Bulgarian perception of these issues, not surprisingly Bulgarians gave much higher priority to issues of security and family and economic well-being than concern for climate change, reporting a relatively low level of knowledge of these issues as well (O'Connor et al. 1999). Thus many recommendations focus on increasing public

---

[*] The authors of this chapter acknowledge the valuable contributions of our colleagues who as individuals and teams prepared the specific analyses for this project. The final list of recommendations, however, is our responsibility, in recognition that not all colleagues would agree on all recommendations. Please refer to the individual chapters to see the way in which contributors framed recommendations from their own viewpoints.

awareness of environmental issues in general (such as drought) and climate change in specific.

We begin with recommendations of highest priority developed from the activities and results of this study. They deal with the explicit challenge of climate change and also with the implicit issues of how Bulgaria prepares itself for climate change impacts, in both scholarly and institutional dimensions. Then, we turn to recommendations from the specific areas addressed in our analyses, drawing upon the observations of our specialist colleagues, while trying to enhance them from the more holistic viewpoint of the study as a whole. Clearly, what we suggest about climate change is also relevant for the more immediate threat of recurring drought.

Recommendations of Highest Priority

The broad goals of highest priority recommendations are to strengthen Bulgarian capability to anticipate and plan for drought and climate change:

1.   *Bulgaria should undertake a comprehensive, integrated national assessment of the potential consequences of climate change.* If there is significant chance that Bulgaria's future *normal* climate could be like conditions experienced in 1993 and 1994, then planning for climate change must have high priority. There is no integrated national assessment that could guide policy- and decision-makers, non-governmental organizations (NGOs), and the public in anticipating and planning for climate change impacts. The assessment should be stakeholder and decision-maker driven, and should be an on-going process, not simply a one-time analysis. By stakeholder and decision-maker driven, we mean that individuals, organizations, firms, and communities affected by the consequences of climate change should help to frame the questions addressed by the scholarly community.

2.   *Bulgaria should learn from contemporary events that help in anticipating the future.* Study of the drought of 1982-1994, and especially its consequences in the 1994-1995 period, shows how analysis of existing problems can identify approaches to managing challenges in the future. We have been unable to address all issues that the drought brought to Bulgaria, but we hope that we have made the case that the drought teaches us how better to anticipate, plan, and react to similar extreme events in the future.

3.   *Bulgaria should ensure that impacts resulting from climate change are incorporated in all aspects of environmental planning at all levels from local to national.* Bulgaria is poised to invest significant capital in infrastructure and

resource development, including plans for major investment in water resources. Whether it is changing climate affecting runoff, drought enhancing the threat of forest fires or challenging agricultural development, floods destroying structures built in inappropriate areas, emergency management during the failure of potable water supply systems, or other aspects of public planning and management, giving attention to future climate could save funds and discourage improvident investment.

4. *Bulgaria must find means to break institutional barriers to sharing of data and to research collaboration.* This study is among few integrated environment-and-society analyses that cross the boundaries of disciplines and institutions, and even here we have failed to adequately exploit expertise that exists in the ministries, university and NGO community beyond the Bulgarian Academy of Sciences. There are significant institutional and financial barriers to the sharing of data and expertise; indeed the trends may be in the opposite direction, of treating information with greater proprietary interest, as a source of needed income, and of treating research opportunity as equally proprietary. It will simply be impossible to address Recommendation 1 without a different attitude toward collaboration and data availability. The mechanisms and spirit of collaboration in this and other global change projects show that overcoming institutional barriers is not impossible (Hristov *et al.* 1999; Staneva *et al.* 2000).

5. *Bulgaria should continue to embrace "no regrets" strategies with regard to climate change.* Implicitly, the Bulgarian commitment to the Kyoto Accords by decreasing the energy intensity of its economy is a "no regrets" strategy that will pay off with greater efficiency and economy, even if its contribution to the global problem is inconsequential compared to larger and richer nations. Planning and action now in anticipation of climate change could prove to be another "no regrets" dimension for the nation, anticipating the extremes of natural variability that will occur without question, even though these conditions might not become the norm.

Climate Change

The broad goals in the field of climate change focus on strengthening understanding of climate change and bringing climate change impacts into the national, regional and local agenda:

6. *Bulgaria should support efforts in relevant institutes in the Bulgarian Academy of Sciences to collaborate with the Scientific Coordination Center for*

Global Change in maintaining national (if not international) leadership in climate change impact research. At the time of writing this book, Bulgaria is among only three nations in the transitional region of Central and Eastern Europe to have a national global change focal point. In spite of this leadership, Bulgaria is weak internally and in its representation internationally in its capabilities for projecting climate change and impacts. A sense of focus and leadership could catalyze substantial accomplishments within existing budget constraints; modest additional resources could have a immeasurable impact through collaboration with international scientific organizations to secure support from funding agencies.

7.   *Bulgaria should provide a program of support to bring climate change issues into the research agenda of the national and regional universities.* These institutions are poised to work more closely with stakeholder and decision-maker communities and help interpret research from the Academy and international sources in terms understandable to local community members as well as to the next generations of public leaders and teachers. In addition, they can be important foci for bringing common concerns from the regions to the national agenda.

8.   *Bulgarian policymakers should encourage issues of climate change causes and impacts to be part of public education from primary through secondary and university education.* There have been considerable strides in bringing contemporary environmental issues into the classroom. These accomplishments should be complemented by using global climate change processes and impacts as further directions for motivating student learning.

Water Resources

The overriding water resource goals are to reduce vulnerability of Bulgaria to drought, whether from climate change or natural variability, and to establish a reliable potable water supply as highest priority during drought:

9.   *A national program for integrated management of water resources, which also takes into account climate change, should be created.* A national water plan has been created, drawing upon extensive Bulgarian expertise on understanding past climate variability. Such planning must now recognize that past climate patterns may be just that, the past. Anticipating climate change is crucial. More effective multiple water use may be a key; structural and management elements to ensure downstream use of hydroelectric water for potable water or other uses are imperative.

10. *A national plan for water management should be devised, including operational plans and structural improvements for the sustainable exploitation of water sources with the aim of overcoming water shortages during drought.* A crucial element of this plan should be the setting of water use priorities during crises.

11. *A long-term program for co-operation between local authorities and NGOs for economical use of water should be developed.* Such activities would include decreasing losses from leaks; introduction of recycling; multiple-use of water, according to its quality; control against illegal drawing of water; and addressing other related issues.

12. *The highest priority for water use during drought should be drinking water.* During times of drought, breach of this rule should carry threat of prosecution when water rationing is introduced, or when instances of water-borne diseases are detected. Legislation must reduce losses so that the sanitary water minimum in dams is guaranteed during time of drought.

13. *Attention must be given to public health dimensions of water supply.* The population should be informed about the quality of water consumed in time of drought, and suitable measures against epidemics should be recommended. Such mechanisms include "boil water" orders to decrease the risk of infection from intermittent water stoppage.

Management of Forest Resources

The strategic task ahead for the forests in the low-lying parts of the country (below 800m) includes adaptation of forests towards a drier climate. For forests in the higher parts of the country (above 800m), the goals of forestry are to preserve biological variety; encourage sustainable development of ecosystems; ensure multifunctional use; and develop a system of protected natural territories:

14. *Local and alien tree and shrub species with proven resistance to semi-arid conditions should be utilized.* Introduce new species from adjacent geographical regions in reforestation work in anticipation of drier conditions.

15. *The importance of a differentiated reforestation in separate regions must be acknowledged.* There is reasonable expectation that climate drying is expected to follow an ascending order: western plains – northern Bulgaria – southern Bulgaria – the Black Sea region – southwestern Bulgaria.

16. *The mixed character of forests should be preserved, with preference for tree species and shrubs which are resistant to climate change.* Multi-layered early-age forest ecosystems appropriate to specific elevations should be created, as such systems are ecologically more resistant to environmental extremes. Preservation and expansion of the mixed character of mountain forest ecosystems is vital.

17. *Secondary (regrowth) forests should be turned into seed forests through the introduction of more resistant and heat-tolerant species.* Special attention should be shown to forest shelter belts, which should be built-up with more drought resistant species and cover larger areas.

18. *Nursery production should provide a gradual change in local and foreign tree and shrub species suited to new conditions.* New technologies are necessary for soil preparation, reforestation, selection measures, in accordance with new ecological conditions.

19. *New methods for selective felling should be devised on the basis of changed ecological conditions and new tree species.* Intensive use of sanitary felling for the quick clearance of forests affected by drought should take place. In higher elevations, contemporary methods of main felling should have natural recovery as a high priority.

20. *An intensive system for the prevention, quick identification, and elimination of threats to forests must be developed.* The threat of forest fires is an obvious area demanding special attention and investment. Creation of a system of pest and disease control under drought is equally important. The development and realization of a protection system against poaching and damage is also needed. Special attention should be drawn to selective felling in young stands to strengthen stand resistance and productivity.

21. *Raise the upper limit of the mountain forests by 100-200m according to changes in climatic conditions.* Nursery production should provide seedling material according to climate change.

22. *Manage mountain catchment areas towards multi-functional operations aiming at higher biological productivity.* Goals include the highest level of water regulation, soil protection, recreation and maintaining the forest microclimate.

23. *Develop and enhance the system of protected areas (national parks, reserves, protected forests) for the preservation of bio-diversity, high-level productivity, and increasing the protective functions of forests.* These areas

provide the "natural controls" for understanding impacts of both climate change and the effectiveness of human managed forests.

Wild Fauna

Goals with regard to wild faunal populations focus on short term assistance for valuable species affected by drought, maintaining or creating diverse ecosystems to support fauna, and anticipating threats from pests:

24. *Use artificial feeding as a short-term response to drought.* In order to reduce fauna losses owing to unfavorable atmospheric conditions it is necessary to assist fauna through artificial feeding followed by initiatives leading to the development of a food base adapted to new conditions.

25. *Activities in improvement of the food base for wild animals should include forest use and management and management of forest meadows and pastures for game.* Feeding of game in the fields requires some managed changes in the landscape, leading to the establishment of shrubbery and other vegetation and watering areas.

26. *Special attention should be given to monitoring populations of harmful animal species and taking preventive measures.* Changing climate may create new opportunities for pest species; early warnings are imperative.

Management of Agricultural Crops

The impacts of drought and climate change on agriculture are among the best-understood areas of analysis in Bulgaria, although existing studies have focused on grain crops, leaving open questions about possible climate change impacts on viticulture, fruits, and vegetable production. The goals here are to maintain and enhance Bulgaria's food security:

27. *A new classification of Bulgarian agro-climatic resources and regions should be developed.* A new classification of regions for agricultural crops, related to new varieties and species which are adapted to a changed climate, would assist private-sector farming decisions in maintaining high productivity. Climate change impacts on viticulture, fruits, and vegetable production should be addressed.

28. *Bulgaria's system of agricultural irrigation should be re-established and made more efficient.* One tenth of Bulgaria's land area was under irrigation in

the 1980s; much of this investment was lost during the 1990s transitional period, facilities that now need to be renovated. Bulgaria must also increase irrigation efficiency through drilling of bore holes; use of new irrigation technology for a reduced flow of water; use of winter rainfall and winter runoff for irrigation; and use of recycled waste water.

29. *Gravity irrigation should be restricted.* Losses under this system are significant. Irrigation should be done in early morning and late evening when evaporative losses would be minimized.

30. *Major attention should be given to reducing water losses from irrigation canals.* All major distribution irrigation canals should be made impervious in order to reduce losses due to infiltration. Permanent canals should be tree-shaded in order to reduce direct evaporation.

31. *Technologies for harvesting melting snow should be used.* This can increase soil moisture storage for perennial plants and wintering crops.

32. *Underground waters in Bulgaria should be used to a greater extent for irrigation.* This strategy could improve the water balance in agriculture as well as save the costs of long-distance water conveyance.

33. *New technologies and efficiencies should be developed to reduce methane emissions from agriculture.* Methane is a significant gas contributing to global warming. Measures to reduce methane emissions coming from biofermen-tation in agriculture and flooded rice should be undertaken, as well as reduction of methane emissions through effective use of manure.

The Economy

The goals related to Bulgaria's economy are related to making the economy more resilient to climate variability and adaptable to potential future climate:

34. *Attention should be given in national economic planning to issues of vulnerability of the economy to climate variability and change.* Sustainable development depends on understanding the environmental limits to sustainability and realistically dealing with these limits in planning national investment. This study found that data systems for understanding drought impacts on the economy were limited; this deficiency needs to be addressed. Bulgaria should provide funding for monitoring and study of the economics of natural hazards. This is especially important for an economy where metrification of such phenomena is particularly difficult. Study must also be undertaken on potential climate change impacts on tourism.

35. *Consideration should be given to the establishment of strategic food and water reserves.* Such a program could be accomplished at the national level, in regional cooperation, or within the broader context of the European Union.

36. *Special notice of issues in the agricultural sector of the economy should be taken.* Pre-accession funds of the European Union should be allocated for measures combating drought, in particular in the field of agriculture. Rapid adoption of agricultural planning and implementation methods for agricultural policy as applied in the EU is recommended, taking into account the status of natural resource endowments.

37. *Understanding of climate change impacts on the energy sector of the economy is critical.* Climate variability and change have major impacts on the availability of hydroelectric resources. A combination of restructuring of the economy and increasing private income may increase electricity demand, even if efficiency of use is increased. Difficult choices lie ahead in Bulgarian decisions for electricity production. How energy would be allocated in the face of diminished hydroelectric production due to drought and to priorities for water use under drought conditions needs to be addressed.

38. *Policy must be developed to deal with conflict of interests between water users in a time of crisis.* Alternatives to allocation may be political (through law), regulatory (through river basin agreements), or through market mechanisms (creating a water market). It is important to anticipate conflicts in water use during drought and to use these potential conflicts both to structure how Bulgaria's water economy will work and to avoid misguided real-time crisis management.

39. *A new approach for forecasting and controlling disasters caused by drought is necessary.* A balanced approach in the centralizing and decentralizing of water use decisions is necessary in order to avoid the multiplication of mistakes.

Health and Hygiene

There can be no higher priority for society than the health of its population. Recommendations related to health and hygiene are based on Bulgaria's need to protect its people from illness related to drought:

40. *Efforts of local government and central authorities to ensure regular and safe water supplies for inhabited areas should continue.* Interruption of water supplies not only has a cost of inconvenience; it also threatens health as

contaminated water infiltrates water systems not under pressure. In addition, lengthy periods without water are a threat to general sanitary conditions.

41. *An overall strategy for ensuring supplies of water for drinking purposes for all consumers should be worked out and maintained.* In order to avoid exhaustion of water sources for drinking water, a local or regional approach for drawing water from separate sources throughout various seasons should be envisaged, taking into account the basic needs of the population and the threat of climatic change.

42. *There must be careful pre-planning of water rationing in anticipation of drought.* Owing to uneven precipitation in the country and instances of local droughts in the absence of reserve water sources, local programs with additional measures for efficient use of drinking water should be prepared. Water rationing plans must include regimes of water use that will prevent reservoir volumes from falling below sanitary minimums.

43. *Health authorities should strengthen and streamline effective anti-epidemic initiatives, especially under conditions of water rationing.* Actions should include conducting a health awareness program, strengthened control of drinking water treatment, "boil water" notices after supply interruptions, control in the processing and preparation of food products, and specific prophylactics for sensitive groups of the population.

Sociology and Ethics of Water

Water has deep roots in Bulgarian society, providing a unique opportunity to build on these traditions in facing water challenges in the future:

44. *Bulgarians should reassess national traditions and search for new approaches for building a new, long-term water culture and water ethic.* These issues should be included in the full educational cycle from an early age to higher education.

45. *Monitoring of social, cultural and aesthetic aspects of water problems in Bulgarian society is necessary.* Water management and use is not simply a matter of economics and policy. The scholarly community, including the social sciences and humanities, has an obligation both to understand and to inform the evolution of social dimensions of this crucial resource. Cooperation and pooling of research efforts in is necessary and can contribute to reasonable decision-making.

The Politics of Water

The goals recommended in view of water politics focus on expanding the public role in water resource and global change decision-making:

46. *The research and artistic communities should be committed to water issues and support the media.* These creative groups can help to influence organizations, the political elite and society as a whole to be aware of and participate in decisions about Bulgaria's environmental future.

47. *Research centers and research organizations such as the Bulgarian Academy of Sciences and universities need specialized units for public relations with the media.* The scholarly community needs to enhance its ability to create interest in the media and to draw the attention of the public to local water problems and national environmental issues.

48. *Government should promote water and global change issues in the media in their variety and breadth.* This supposes both a commitment to stakeholder involvement with decision-making, with journalists providing informed discussion of water problems.

49. *The media and public should be provided with accessible, non-technical information from scientific research.* Institutions such as the Bulgarian Academy of Sciences should produce literature for the general public.

50. *The Bulgarian community can learn from the conflicts occurring during the drought about broader dimensions of how it wants to address issues of local versus national, and decentralized versus centralized management of critical resources.* Events such as the drought reported in this book point to important questions about how Bulgaria will balance decisions that give priority to environmental quality, economic development, and emergency preparedness. The drought also pointed to issues of fairness, equity, and transparency that will always pervade critical public decision-making.

Conclusions

The drought of 1982-1994, culminating in the water emergency of 1994-1995, is both an analog for the challenges that Bulgaria may face with climate change and at the same time a parable from which much can be learned about society-environment relations during a time of crisis. The foregoing recommendations encapsulate our observations about how we might turn what we have learned toward improving Bulgaria's future. We trust that at least some of these recommendations will become reality.

References

Energoproekt *et al.* 1999. *Bulgaria Country Study to Address Climate Change: Final Report.* Sofia, for U. S. Country Studies Program.

Hristov, T. *et al.* 1999. Глобалните промени в България. София, Национален координационен център по глобални промени.

National Climate Change Action Plan. 1999. Bulgarian Ministry of the Environment, Energoproekt, Sofia.

O'Connor, R. E., R. J. Bord, A. Fisher, M. Staneva, V. Kozhouharova-Zhivkova, and S. Dobreva. 1999. "Determinants of Support for Climate Change Policies in Bulgaria and the USA," *Risk Decision and Policy* 4(1):1-16.

Staneva, M. P., C. G. Knight, T. N. Hristov, D. Mishev (editors) 2000. *Global Change and Bulgaria.* Sofia: National Coordination Center for Global Change and University Park: Center for Integrated Regional Assessment.

Note on Transliteration

The Bulgaria Cyrillic alphabet is transliterated into the Roman alphabet using several alternative conventions, including those based on English, French or German. In this book, the editors have elected to use English-based transliteration where possible, while respecting traditionally established patterns of transliteration, as well as preferences of authors, when known.

Examples of alternative transliterations include the following:

Bulgarian Cyrillic	English	German	French
Боровец (resort)	Borovets, Borovetz	Borovets, Borovetz	Borovets, Borovetz
Бургас (city)	Burgas	Burgas	Bourgas
Гешев (name)	Geshev	Geshew	Gechev
Искър (River)	Iskar, Iskur, Isker	Iskar, Isker	Iskar
Кюстендил (city)	Kyustendil	Küstendil	Kyustendil
Родопи (mountains)	Rhodope, Rhodopi or Rodopi	Rhodope, Rhodopi or Rodopi	Rhodope, Rhodopi or Rodopi
Русе (city)	Russe or Ruse	Ruse	Roussé
София (city)	Sofia	Sofija	Sofiya
Странджа (mountain)	Strandja or Strandzha	Strandja	Strandja
Христов (name)	Hristov or Christov	Kristov	Christoff or Christov
Янкова (name)	Yankova	Jankova or Jankowa	Iankova
Янтра (river)	Yantra	Jantra	Iantra

The transliterations used in this book do not necessarily have official status. The editors apologize for any oversights in this regard.

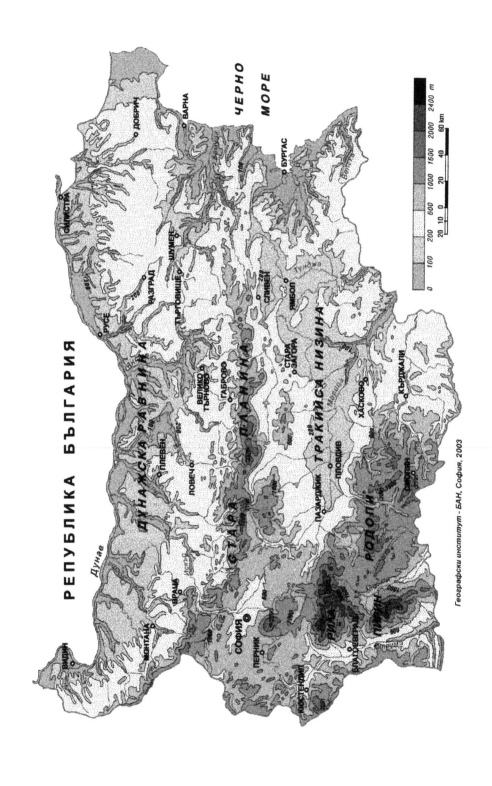

РЕПУБЛИКА БЪЛГАРИЯ

ЧЕРНО МОРЕ

ДУНАВСКА РАВНИНА

СТАРА ПЛАНИНА

ТРАКИЙСКА НИЗИНА

РОДОПИ

РИЛА

ПИРИН

Дунав

ВИДИН

МОНТАНА

ВРАЦА

СОФИЯ

ПЕРНИК

КЮСТЕНДИЛ

БЛАГОЕВГРАД

ЛОВЕЧ

ПЛЕВЕН

ВЕЛИКО ТЪРНОВО

ГАБРОВО

ПАЗАРДЖИК

ПЛОВДИВ

СТАРА ЗАГОРА

ХАСКОВО

КЪРДЖАЛИ

СЛИВЕН

ЯМБОЛ

БУРГАС

ВАРНА

ДОБРИЧ

СИЛИСТРА

РУСЕ

РАЗГРАД

ТЪРГОВИЩЕ

ШУМЕН

2400 m
2000
1600
1000
600
200
100
0

60 km
40
20
0

20 10 0

Географски институт - БАН, София, 2003

Index